FUNDAMENTALS OF
CONFIGURATION MANAGEMENT

FUNDAMENTALS OF CONFIGURATION MANAGEMENT

THOMAS T. SAMARAS
Heuristics, Inc.

FRANK L. CZERWINSKI
TRW Systems Group, TRW Inc.

WILEY-INTERSCIENCE

A Division of John Wiley & Sons, Inc.

New York · London · Sydney · Toronto

Library of Congress Catalog Card Number: 75-127668

ISBN 0-471-75100-6

Printed in the United States of America

10 9 8 7 6 5 4 3 2 1

- **TO MY WIFE MARY SAMARAS AND TO BILLY AND DANNY JONES**

- **TO MY WIFE AND FAMILY ANN, LYNNE, AND MICHAEL CZERWINSKI**

ACKNOWLEDGMENT

We wish to express our appreciation to Edward G. Hantz and Alan E. Lager of the Grumman Aerospace Corporation, who have allowed us to use much of the content of their presentation as a basis for "setting the stage" in Chapter 1, to D. W. Michler, TRW Systems, Software and Information Systems Division, for his advice, consultation, and "real-world" inputs during preparation of Chapters 20 and 21, and to Walker Bennett, Huntsville, Alabama, for his general comments on the entire manuscript. Special thanks go to Professor Paul Zall of California State College at Los Angeles whose teaching inspired the writing of this book.

T.T.S.
F.L.C.

PREFACE

During the last 15 years aerospace programs have emphasized areas of the engineering profession previously unknown to many engineers and managers. Configuration management is one of these areas that have become increasingly important to programs involving the design, development, production, testing, and deployment of aerospace instruments and systems. To many engineers and managers configuration management is an esoteric term, but one that they are encountering more frequently in their daily routines because of the government's growing demands for stricter control over the equipment it buys. Configuration management is the discipline of ensuring that equipment or hardware meets carefully defined functional, mechanical, and electrical requirements and that any changes in these requirements are rigidly controlled, carefully identified, and accurately recorded.

Because of the importance and wide application of configuration management in space and atmosphere instrumentation work, most aerospace engineers become involved in it to some degree and therefore must be familiar with its techniques and principles to perform their jobs effectively. Indeed, many government contracts explicitly require that configuration management be employed on projects they cover. Thus technical people who work in equipment design, manufacturing, quality assurance, testing, documentation, or management must have a basic knowledge of configuration management objectives and procedures.

The purpose of this book is to describe the fundamentals of configuration management so that engineering students, designers, engineers, project managers, and administrative personnel can have a common language for working effectively with their configuration management counterparts—for it is certain that during their careers in the aerospace industry they will require an understanding of the configuration manager's role and activities on their projects.

Configuration managers will find the information presented useful as a reference for basic configuration principles and as a guide for their assistants and fledgling configuration managers. This book will also provide engineers interested in good management with a broad view of engineering control activities related to the successful completion of a project.

September, 1970
Los Angeles, California

THOMAS T. SAMARAS
FRANK L. CZERWINSKI

ix

CONTENTS

FUNDAMENTALS OF
CONFIGURATION MANAGEMENT

Chapter 1

INTRODUCTION

Although "configuration management" is an esoteric term, its concepts and practices are universally applied by industry to deliver a product that meets customer requirements. Of course, the degree and consistency of application vary widely, depending on the complexity and use of the product and the management philosophy of the company. For sophisticated products built to stringent standards such as those imposed by the government for aerospace and military systems, the formalized discipline of providing uniform product descriptions, status records and reports, and change control is called configuration management. The formal implementation of this discipline is a major task involving numerous specialized concepts, personnel, and procedures. This book describes the fundamentals, techniques, and latest developments of configuration management so that its basic principles can be rationally and effectively applied.

Chapter 1 is devoted to general configuration management topics, such as basic concepts, definitions, objectives, and techniques, as well as its history and its future. The following chapters describe the detailed aspects of configuration control, identification, and accounting. Because of the specialized terminology, a list of acronyms and a detailed glossary are included for reference.

The term "product" is used in this text to represent a system,[1] subsystem, equipment, instrument, component (power supply, regulator, etc.), data software, or operational computer software. The official term applied to the total product or selected portions of the product is configuration item (CI).[2] Although "contract end item" (CEI) is the more traditional term, all elements (end items) of a product are not necessarily line items of a contract; therefore, CI is the preferred term and will be used throughout

---

[1] A system is a grouping of subsystems, equipments, and components. The term may also include personnel and facilities.

[2] The selection of CI levels comprising a parent product is discussed in Chapter 7.

the following chapters. In either case, CI and CEI identify a product, or product element, as being subject to configuration management procedures and requirements.

1.1   BASIC CONCEPTS

With thousands of engineering changes a normal part of the development of a complex product, how can the customer and builder be sure that the product initially described in a specification and contract is what is delivered? Stripped to essentials, configuration management directs itself at this need to know just what the product, operational computer software,[3] and sup-supporting data (drawings, operating manuals, repair instructions, etc.) consist of as the project evolves with changes that occur daily.

"Configuration management implicitly accepts these changes during the progress of a project as something normal—not something to be deplored, but something to be managed. This means that, by the process of configuration management, we progressively define the shape and form of the project or item we are managing. We must accept, as a necessary condition for technological growth and cost effectiveness,[4] the idea that change when properly managed is a vehicle for optimizing costs and other interface[5] and project control considerations, but in the absence of control of change we have, in effect, technological anarchy."[6]

Configuration management is really the formalization of a discipline which every effective manager adapts informally in one way or another. It integrates the technical and administrative actions of identifying, documenting, controlling, and reporting the functional and physical characteristics of a product during its life cycle; it also controls changes proposed to these characteristics. Thus configuration management is the means through which the integrity and continuity of technical and cost decisions related to product performance, producibility,[7] operation, and maintenance are recorded, communicated, and controlled by project managers.

[3] Computer instructions functioning as operational entities or in conjunction with associated hardware in achieving system requirements.

[4] Cost effectiveness is the ability of an item to fulfill a specific need at minimum cost during its life cycle.

[5] Interfaces are common boundaries between two or more products, companies, or government agencies where physical and functional compatibility are required.

[6] Presentation by Mr. George E. Fouch, Deputy Assistant Secretary of Defense (Installation and Logistics) Management Systems and Programs, before the Third Annual Configuration Management Workshop of the Electronics Industries Association, San Diego, California, November 19, 1968.

[7] Ease of manufacture and assembly of an item, including access to its parts, tooling requirements, and realistic tolerances.

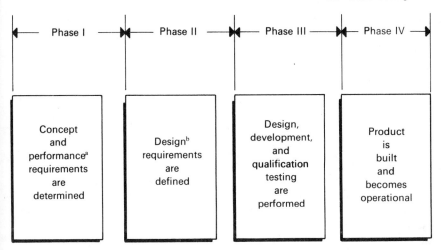

[a] The meaning of "performance" depends on the product type; e.g., the performance of an aircraft would be defined as its maximum range, maximum and cruising speed requirements, airlift capacity, etc.

[b] Design requirements define the product's mechanical and electrical characteristics; e.g., its weight, size, power consumption, center of gravity, operating temperature, etc.

Figure 1.1 Life cycle of a product.

The term "progressive definitization" is a key concept in configuration management that should be stressed in the following respect: as a product proceeds through research, conceptual design, development, detail design, qualification, first article (product) production, and follow-on production, the identification of its configuration becomes progressively more definite and precise, and may be continually modified throughout the service life of the product. Accordingly, the configuration of a product is *derived* during development, *determined* during design, *established* during production, and *maintained* during operational support. Operational support refers to the requirements for operation, maintenance, and repair of a product during actual use.

Figure 1.1 illustrates the phases of a product's life. Regardless of the titles you choose to put on these phases, they always exist within good engineering and business practice in industry.

Any product, be it an aircraft or a pencil, must have some reason for being, a function that has to be performed—a performance requirement. While this requirement and the general form of the product are being established we are in the initial phase: *concept and performance requirements determination.* As soon as a limiting set of restrictions is placed on the product, such as size, weight, and durability, the second discernible phase has been entered—the *design requirements* have been defined. In the third phase, *development*, the physical characteristics of the product are defined in sufficient detail to enable it to be built and tested to see if it fulfills its intended functional (or performance) requirements. The final phase, *operational*, is the period in which the CI is produced and serves its intended function.

In a buyer-seller relationship, particularly in the aerospace marketplace where the buyer is usually purchasing not only the end product but its design and development as well, the point of departure or "baseline" between each of these phases has significance. Each represents a point of decision by the buyer or negotiation between the buyer and seller, or both. The buyer must have some measure of supervision and control over the seller's activities to assure that (a) he has sufficient basis for making the basic critical decisions, such as continuation, cancellation, or modification of a project; (b) he is getting the product contracted for at all times; and (c) the product will be compatible with the other CI's in his complement of equipment or associated project interfaces.[8]

It has always been difficult to determine the optimum measure of control that should be applied. The cost and effectiveness of this control vary considerably with the product's life cycle. If the variation is to be in definable degrees, definitive baselines between the phases of the product's life cycle must be established. Then the seller's flexibility for creativeness and ingenuity can be greatest during the early project phases where it will be most beneficial, and the buyer's control can become greater in later phases as the cost of a change becomes greater. The respective emphasis and considerations during the baseline phases are briefly described in this chapter. A detailed examination of a typical product cycle and the application of baselines within the phases of a product's life are provided in Chapter 20.

1.2 FOUR CRITICAL DEFINITIONS

A clear understanding of the concepts and techniques embodied by the terms "configuration management," "control," "identification," and "accounting" is essential to understanding the following text. Therefore, we present definitions of these four terms before discussing other subjects. Note that control, identification, and accounting are of equal importance to configuration management and the order of presentation does not represent an attempt to emphasize one topic over another.

[8] Training and test programs, personnel, etc.

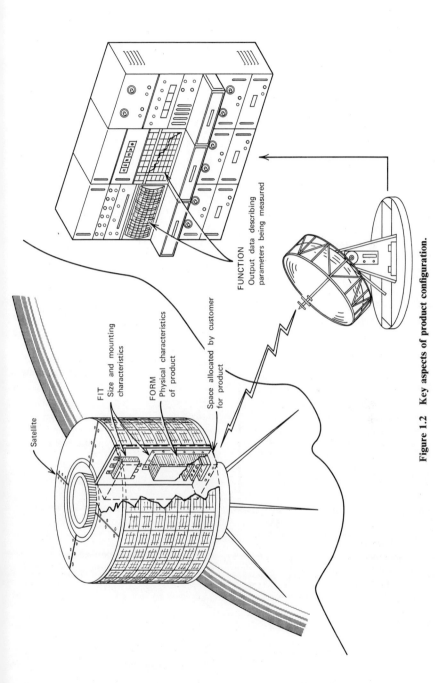

FIT
Size and mounting
characteristics

FORM
Physical characteristics
of product

Satellite

Space allocated by customer
for product

FUNCTION
Output data describing
parameters being measured

Figure 1.2 Key aspects of product configuration.

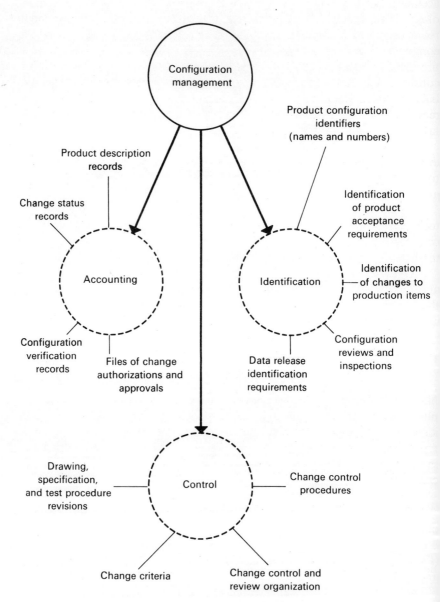

Figure 1.3 Major facets of configuration management.

Configuration Management. Configuration management is the art of organizing and controlling planning, design, development, and hardware operations by means of uniform configuration control, identification, and accounting of a product. The goal of these operations, in colloquial terminology, is to assure that the delivered CI meets form, fit, and functional requirements as simply illustrated in Figure 1.2. "Configuration" refers to a complete description of the physical and functional characteristics of a product; for example, its weight, shape, size, materials, processes, power consumption, and performance (measurement range, accuracy, stability, linearity, etc.). The term configuration also applies to technical descriptions required to build, test, operate, and repair a CI. Figure 1.3 shows the major facets of configuration management.

Configuration Control. Configuration control involves the systematic evaluation, coordination, and approval or disapproval of proposed changes to the design and construction of a CI whose configuration has been formally approved internally by the company or by the buyer, or both.

Configuration Identification. Configuration identification refers to the technical documentation that identifies and describes the approved product configuration throughout the design, development, test, and production tasks. It also applies to the identification of changes and to product markings.

Configuration Accounting. Configuration accounting is the recording and reporting of CI descriptions and all departures planned or made from the CI through the comparison of authorized design data and the fabricated and tested configuration of the CI.

1.3   OBJECTIVE

The overall objective of configuration management is to guarantee the buyer that a given product is what it was intended to be—functionally and physically, as defined by contractual drawings and specifications[9]—and to identify the configuration to the lowest level of assembly required to assure repeatable performance, quality, and reliability in future products of the same type. To satisfy this objective the following five major goals are commonly an integral part of the configuration management effort:

1. Definition of all documentation required for product design, fabrication, and test.
2. Correct and complete descriptions of the approved configuration. (Descriptions include drawings, parts lists, specifications, test procedures, and operating manuals.)

---

[9] A specification is a document that defines the explicit requirements for a product.

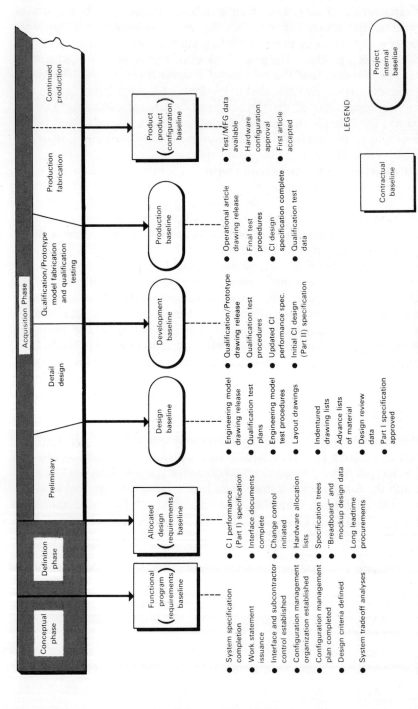

Figure 1.4 Typical baselines.

3. Traceability of the resultant product and its parts to their descriptions.

4. Accurate and complete identification of each material, part, sub-assembly, and assembly that goes into the product.

5. Accurate and complete pre-evaluation control and accounting of all changes to product descriptions and to the product itself.

Achievement of the above goals will ensure a successful configuration management program. The quantity of data that must be identified, controlled, and accounted for, however, can present a major challenge to even the most talented administrators and managers. Therefore the small number of goals in no way reflects the magnitude or complexity of configuration management activities.

1.4 BASELINE MANAGEMENT

Figure 1.4 illustrates the "baseline management" concept that forms the framework for "progressive definition," described in a preceding section, and the necessary commensurate control points for assuring satisfactory definition of the product. Baseline management covers three phases, usually called conceptual, definition, and acquisition, and six baseline points, three of which are contractual and three of which are suggested as typical internal project baselines. The initial contractual baseline—functional or program requirements—is established by performance and objectives criteria documentation that consist of specifications defining the program requirements. Once these specifications have been approved, any changes to them must also be approved.

When the functional requirements baseline has been further allocated to various elements (configuration items) of the product and their design requirements have been defined by means of the next lower tier of specifications (which are usually performance specifications for individual CI's comprising the product), the allocated (or design requirements) baseline is established and approved by the buyer. Once this step has been taken, the buyer must then approve all changes affecting specifications defining both program and "design to" requirements for the product or its configuration items respectively.

We now progress through a series of internal baselines (design, development, and production) into the operational design and prototype or qualification model fabrication and test portion of the acquisition phase (that is, the period between end of definition phase and delivery of the last product to the customer).

As shown in Figure 1.4, the design baseline is represented by the full complement of engineering model drawings, preliminary engineering drawings, and supportive information for the qualification, prototype, and operational articles.[10] The qualification model and prototype drawing release data, updated performance specification, initial design specifications, test plans, and test procedures form the development baseline. The third baseline, the production baseline, is the operational article drawing package ("prototype-update"); final specifications; parts, process, and materials documents; final test procedures, and so on.

The next level of configuration baseline authority now consists of design documentation that for clarity we will refer to as "build-to" specifications, engineering drawings, manufacturing and quality data, test procedures, and other detailed technical documentation. It is to these data that the product (or product configuration) baseline is established. The most formal phase of control is begun at this time. The buyer now requires that he approve all changes affecting the configuration to the extent specified in the contract. (A contract is a legal agreement between the buyer and seller defining the work to be done and the products to be delivered.) A few areas that would require change approval include:

1. Performance.
2. Weight.
3. Interfaces with other CI's.
4. Contract price.
5. Delivery schedule.
6. Operator safety.
7. Reliability.

Figure 1.4 shows six baselines and the documents and tasks that represent their satisfactory achievement. Note that the term "program" is used to indicate an effort that encompasses a number of smaller efforts called "projects"; for example, the *Apollo* program consists of many projects, such as the solar wind spectrometer design and development project, and a program specification in this case establishes the total requirements for a related series of projects designed to accomplish a broad scientific or technical goal.

1.5 CONFIGURATION MANAGEMENT TECHNIQUES

As mentioned before, configuration management applies technical and administrative direction and surveillance to configuration baselines and increased control as each baseline is established. The basic techniques,

---

[10] A qualification article is used to demonstrate that the CI is capable of performing satisfactorily in the worst environment expected. A prototype is the first production model that meets all customer requirements; it is usually not flown. The operational model is a production CI that is used to satisfy the customer's needs.

methods, or procedures that enable the discipline to be systematically applied are illustrated in Figure 1.5 and discussed in the following sections.

Configuration Identification

Configuration identification consists of creating and formally releasing the complete technical documentation, including specifications, drawings, and data lists, which defines the original approved configuration (baseline) and subsequently defines all approved changes to the individual CI's. It should also be understood that this involves physical identification of parts, subassemblies, assemblies, and facilities as specified in drawings and specifications. It also involves the identification of the design and as-built documents themselves so that they may be readily associated with the CI's they support. Identification therefore requires systematic control of part, specification, and data list numbers as well as the assignment of serial numbers. (Serial numbers enable identification of individual items that are otherwise identical.) Configuration identification assures that hardware and all supporting documentation are continually compatible for the life of the CI.

Configuration Control

Configuration control is a continuing function, beginning in the earliest stages of a project and extending over the full service life of the individual CI. It consists of those systematic procedures by which configuration changes are proposed, evaluated, coordinated, and either approved for incorporation or disapproved. Initial phases of configuration control, before buyer acceptance of the first CI, are directed toward control of configuration as defined in documents, primarily specifications. Following first CI acceptance, configuration control is focused primarily on hardware with changes to documentation occurring as a result of approved specification and hardware changes.

Configuration Verification

Having identified the approved product configuration and having controlled changes to the configuration up through the point of authorized release, it is then necessary to verify the physical incorporation of changes. Configuration verification involves continuous comparison of the changed hardware with the engineering data that define its approved configuration. Inherent in this verification process is the discovery and correction of configuration discrepancies by (a) rework of the CI to its approved configuration or (b) approval by the buyer of the CI as-built configuration, using officially prescribed procedures and forms to record the approval.

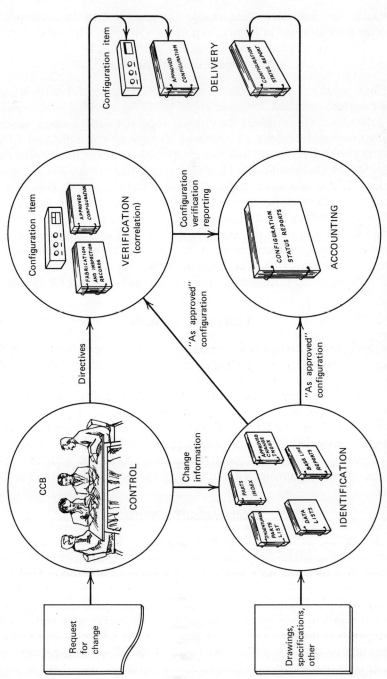

Figure 1.5 Configuration management functions.

The end product of configuration verification is a CI that is perfectly mated to all supporting technical documentation or, failing this, that has all configuration discrepancies identified and specifically approved by the buyer.

Configuration Accounting and Reporting

Having accomplished the functions of configuration identification, control, and verification, we then make the previous tasks meaningful by the processes of configuration accounting for internal purposes and of configuration reporting for external (buyer) purposes. This involves the creation of documentation, in formats specified by contract, which is used internally and then reported to the buyer to provide him with accurate information on the configuration status of all CI's entering his inventory or stock area. This reporting activity continues during the operational life of the CI, continuously reflecting changes authorized for factory or field incorporation.

In summary, the configuration status accounting technique establishes records which enable proper logistics support[11] to be established. These records include (a) where a product is located or installed; (b) the identification of selected product items by serial number makeup, and (c) current modification status. The complexities of these records must be consistent with the configuration identification and must be established by the buyer on a case-by-case basis to suit the required level of indenture control and intended use.

Configuration Audit and Review

To ensure that the physical and functional characteristics of a CI match those specified in the product's identification, formal configuration audits are normally performed on selected items in incremental progression leading up to establishment and validation of a product baseline. These audits are called functional configuration audit (FCA), physical configuration audit (PCA), first article configuration inspection (FACI), first article configuration review (FACR), configuration acceptance inspection (CAI), or configuration audit review (CAR), depending upon the service, agency, or "fashion" at the time the contract is signed.

1.6   KEY FEATURES OF SUCCESSFUL CONFIGURATION MANAGEMENT

As in any field of administration and management, certain characteristics or qualities are particularly important to successful configuration management. "Successful" in relation to configuration management means that

---

[11] Project area responsible for supplying spares, instruction manuals, training and maintenance facilities, and product modification kits.

the five goals previously listed under *Objective* are satisfied at minimum cost to the buyer and seller. The key features that characterize successful configuration control, identification, and accounting are the following:

1. *Early* and *complete* definition of configuration management goals, scope, and procedures.
2. *Speed* in evaluating and processing changes.
3. *Accurate* identification and accounting of changes.
4. *Complete* descriptions of changes.
5. *Close coordination* among key elements of the project team.
6. *Cooperative* and *responsive* buyer.
7. *Minimum* labor requirements.

Two other important features are (a) development of the simplest configuration control, identification, and accounting approach or system that will provide the desired results and (b) minimum number of forms and related documents for implementing changes and for providing complete records of all changes.

1.7 QUALITIES OF A CONFIGURATION MANAGER

Since the configuration manager is the key project member for assuring that a product meets the buyer's requirements, we present a brief description of his necessary qualities.

An exact and all-inclusive identification of the qualities that make up a good configuration manager cannot be provided because of the variable requirements of different projects and companies. However, certain general qualities can be given as a guide for selecting prospective configuration managers. The correct mixture of qualities that will provide the most effective manager will depend on the nature of the CI and on configuration requirements applicable to the project. With these considerations in mind, the qualities of a configuration manager should be the following:

1. Administrative and organization ability.
2. Aggressiveness with diplomacy and tact.
3. Technical knowledge (design, manufacturing, quality control, and testing).
4. Documentation skills.
5. Attention to details.
6. Familiarity with the company.
7. Familiarity with the product.

In particular, he must be able to focus on the overall picture of a project and yet not ignore small details; he must be able to plan the configuration management system and to coordinate the work of people in widely different activities; and he must be capable of resisting demands for short cuts when an emergency arises.

1.8 HISTORY OF CONFIGURATION MANAGEMENT

A good part of the techniques and formalization of configuration management is based on good business practices, but, regardless of the self-imposed discipline practiced by companies, both the depth and uniformity of documentation relating to configuration control and accounting, historically, have been comparatively shallow.

This deficiency became apparent in the race for a successful missile launch in the 1950s. With time being critical, the promulgation of changes was accelerated to resolve incompatibilities among elements supplied by many associate and supporting contractors.[12] When a successful flight was finally made and the buyer, in the euphoria of success, said, "Build me another one," industry found themselves the following circumstances:

1. Their prototype was expended (launched into trajectory).

2. They did not have adequate records of part number identification, chronology of changes, nor change accomplishment. Technical publications did not reflect all the various changes that had been made, and it was obvious that a second success could not be guaranteed, nor an identical article produced.

This, of course, is gross oversimplification, but with the era of manned space flight approaching, this situation had to be improved.

Major General William B. Bunker of the U.S. Army Materiel Command described the situation in these terms:

"Expensive and large systems were being delivered with no information as to how they were built and how to keep them running. Deficiencies and failures in tests started long sleuthing operations to determine what had failed and what design alternatives were available to solve the problem. Illustrated parts lists appeared months, sometimes even years after the equipment [was delivered], and were usually inaccurate and incomplete. We qualified certain components and tests and then found, usually after they failed in service, that the production design had been altered, and we found that only the original manufacturer could supply the system or its component because we couldn't tell anyone else how it was made."[13]

---

[12] Same as seller or company.

[13] Speech given by Seymour J. Larber for Major General William B. Bunker, Deputy Commanding General, United States Army Materiel Command, to the American Society for Quality Control, November 18, 1965.

Aircraft weapon systems were becoming quite complex. Military aircraft carried sophisticated electronic hardware which required on-line maintenance, and air bases and aircraft carriers now devoted great proportions of their space to electronic and avionic (electronic components used on aircraft) support facilities. Multiple configurations of CI's were often undiscovered until maintenance, troubleshooting, spares interchangeability, and supporting documentation presented compatibility problems.

During this period the services were able to initiate a new weapon systems program based on conceptual studies that demonstrated only that a militarily useful system could be developed. The costs of development and production were accounted for separately, from the costs of operation. A point was reached in 1957, for example, when one service found that more weapon systems had been authorized than they could carry the cost for.

The "formal set of procedures" establishing configuration management have in the main, grown out of military specifications and contract exhibits. The mad race for arms during World War II taught some bitter lessons. Engineering changes levied on increasingly sophisticated systems too often led to products that turned out to be a surprise.

The single most effective postwar regulation that took change activity out of the conversation stage was *ANA Bulletin* (Army, Navy, and Air Force) No. 390. This established the "engineering change proposal" (ECP) and gave industry uniform guidelines for proposing aircraft changes.

Many other contract exhibits dealing with the subject followed, but the next most significant one was *ANA Bulletin* No. 391a. This further broke down ECP's by classification of priority, and it extended the discipline beyond the airframe industry into electronics and manufacturers of ground support equipment. This instruction played a significant part in the development of supersonic aircraft and ballistic missiles.

ANA Bulletin No. 445 was completed in 1963. It combined and superseded *ANA Bulletins* 390 and 391a. It was a distinct improvement over both for it provided a uniform procedure for the submission of proposed engineering changes to the government for approval. *ANA Bulletin* No. 445 reiterated the necessity for including the technical, fiscal, and logistic supporting information required to define the impact of a proposed engineering change on the management of a program. Over and above the earlier bulletins, it added maintainability and reliability as elements requiring consideration as Class I changes.

ANA Bulletin No. 445 has presently been superseded by MIL-STD-480. MIL-STD-480, "Configuration Control—Engineering Changes, Deviations, and Waivers," represents the most complete description of change control. This standard and other related documents are described in the following sections.

Life-Cycle Costing

Awareness of these problems spurred studies of life-cycle costs of products and related analysis of their cost effectiveness.[14] Essentially, these studies recognized that (a) the initial development and acquisition cost of a product was but a small portion of the total cost of ownership; (b) cost comparisons should be related directly to the capability for performing assigned missions; and (c) the practices of evaluating proposals for new programs on the basis of development and production costs alone required reevaluation.

In fact, many fundamental changes took place in the character of the aerospace industry during the early 1960s. Among them, mission-oriented, five-year force structure[15] and financial plans came into being. As a result, the services are now required to make system effectiveness,[16] cost effectiveness, and trade-off studies to establish the best solutions to meet national defense objectives. To gain authorization to modify the five-year force plan program change proposals must be submitted to DOD (Department of Defense). In essence, both a contract and configuration management interface exist between DOD and each service. For each change to the force structure, a preliminary technical development plan (TDP) is created which establishes a baseline for program requirements.

Management Systems Developments

While these developments were progressing at DOD, the Air Force was developing techniques of configuration management to be applied to contractors. After exhaustive research, industry surveys, management consultation, and large expenditures, the AFSCM 375 Systems Management Series was born. This series of management volumes provides step-by-step instructions for systems management by program office directors in the complicated business of bringing sophisticated products from an initial concept to an operational status in the field.

Industry got a hurried look at a working draft of AFSCM 375-1, "Configuration Management During Development and Acquisition Phases," at the Configuration Symposium in Los Angeles in February 1962. The 19 exhibits of AFSCM 375-1 were, in effect, a series of specifications laid end-to-end in an effort to give a complete resource management procedure from definition to acquisition.[17]

[14] Cost effectiveness is the ability of an item to fulfill a specific need at minimum cost during its life cycle.

[15] The 5-year force structure establishes current and "packaged" military requirements covering a 5-year period for DOD and congressional review of government programs.

[16] System effectiveness is the degree to which a system can be expected to achieve specific mission requirements. Effectiveness is measured in terms of system availability, dependability, and capability.

[17] Lt. Col. Ben N. Bellis, *Aerospace Management Publication*, April 1964.

In June 1962 AFSCM 375-1, "Configuration Management During the Development and Acquisition Phases," was released for publication. This document established a new concept of specification management and initiated the application of baseline concepts, configuration record requirements, seller-buyer interface responsibilities, and an elaboration on the change management requirements of *ANA Bulletin* 391a. AFSCM 375-1 defined the program requirements baseline (now called functional baseline), and the product configuration baseline (today identified as the production baseline). In November 1963 the Air Force revised Air Force Regulation 375-1 to incorporate the definition phase and to establish a middle baseline called the design requirements baseline, which is known today as an optional baseline called allocated baseline.

In June 1964 the AFSCM 375-1 was extensively revised to include the definition and acquisition phases. This manual is based primarily on three source documents: *Defense Standardization Manual* M-200A, Military Specification MIL-D-70327, "Drawings and Data Lists," and *ANA Bulletin* No. 445, "Engineering Changes." It superseded about 61 government specifications, bulletins, and standards. In parallel, the Department of Defense during February 1964 released DOD Directive 3200.9 covering the requirements for concept formulation and contract definition. Since that time, two iterations of DOD Directive 3200.9 have taken place and the definition phase is an established fact today.

Proliferation and Problems

The National Aeronautics and Space Administration (NASA), recognizing the importance of configuration management, published its own document on May 18, 1964, which is entitled *Apollo Configuration Management Manual*, NPC 500-1, and closely resembles AFSCM 375-1. However, the application of this manual to the *Apollo* program in midstream required an extensive supplement. In its general principles, the NASA manual is similar to the Air Force document, but small differences in terminology and shading of definitions exist.

A year later the Army Materiel Command issued its configuration management regulation, AMCR 11-26; while similar in many respects to AFSCM 375-1, it lacks the extensive detail and definitions of terms. Further similarity in principle was evidenced when the Army released AMCR 11-16, "Total Decision Making Process." These two documents are the Army's equivalent to the Air Force's AFSCM 375-1 and comprise a total integrated system for managing selected systems within the Army Materiel Command.

The Navy provided detailed instruction for processing of engineering change proposals in BuWeps Instructions 5200.20 and *ANA Bulletin*

No. 445, Supplement No. 1; Systems Management 5200.11; and NAVAIRINST 4000.15, "Management of Technical Data and Information." The Navy Materiel Command and the Antisubmarine Warfare Special Project Office issued brief configuration management documents for these particular projects.

A chronology of configuration management developments from 1962 to 1967 follows:

Date	Agency	Document Number
Jan. 1962	Air Force	AFSCM 375-1
Nov. 1963	Air Force	AFR 375-1
May 1964	NASA	NPC 500-1 (now NHB 8040.2)
June 1964	Air Force	AFSCM 375-1 (Revised)
Dec. 1965	NASA	Manned Spacecraft Center (MSC) Supplement to NPC 500-1
Jan. 1965	Army	AMCR 11-26
Dec. 1965	Navy	ASWSPO 5200.4
Feb. 1966	Navy	NAVMAT INSTR 5000.6
Sept. 1967	Navy	NAVMAT INSTR 4130.1

The proliferation of these documents and the placing of them on contracts were not without opposition from industry nor without problems. As with other management concepts embraced by the agencies of DOD and NASA as well as other government agencies, configuration management was adopted gradually over a period of a few years. As an agency became configuration management oriented, it issued a new implementary document covering and defining new terms with slight variations from those appearing in preceding documents. The frustrating result has been that although each agency has achieved a working system suited to its particular needs, major subcontractors who deal with multiple prime contractors must now contend with multiple requirements. They are similar in concept but encompass a multitude of variants in the details of significant tasks and reporting requirements.

Attempts at Standardization

Extensive efforts have been made by the government in concert with industry to reduce the proliferation of configuration management instructions. In 1968 all service agencies were given directions by DOD to avoid releasing any additional implementation documents defining configuration management procedures, policies, and so forth, pending issuance of DOD directives, instructions, and standards.

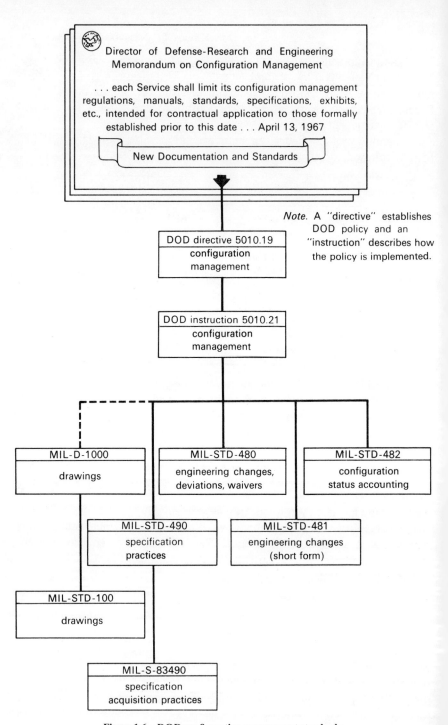

Figure 1.6   DOD configuration management standards.

The Department of Defense's approach to configuration management is to require the optimum degree of uniformity in regard to policy, procedures, data forms, and reports at all interfaces within the DOD and between DOD and industry. To achieve this goal, DOD policy and guidance have been supplemented by a limited number of mandatory, coordinated military standards and specifications for contractual application in lieu of the variety of existing exhibits and other documentation used for this purpose. Although DOD was the prime mover towards standardization, the National Aeronautics and Space Administration, representing a nonmilitary-oriented R & D agency of the Government, contributed to this effort and was an advisory member to the DOD team responsible for establishing uniform procedures and documents.

The DOD standardization efforts in the field of configuration management have been long awaited by industry. Recently final versions of drafts that have received extensive review by industry over a period of eight or nine months have been released. These standards are listed in Figure 1.6. All these documents have been previously released in draft form for review through various industry groups including the Defense Industry Advisory Committee (DIAC), Council of Defense and Space Industries Associations (CODSIA), and its member associations, such as Aerospace Industries Association (AIA) and Electronic Industries Association (EIA).

Government Directives and Instructions

Since the government has pioneered the development and application of formalized configuration management techniques, it is natural that it has also taken the lead in preparing directives and instructions for uniform application of configuration management procedures. On July 17, 1968, with the issuance of DOD Directive (DODD) 5010.19, the government began this process, which will take several years to apply. This directive establishes basic DOD policy and criteria governing the configuration management of defense material items.

The Directive applies to all echelons and all government-industry interfaces. The document clearly states that configuration management application shall be carefully tailored to be consistent with the quantity, size, scope, stage of life cycle, nature, and complexity of the product involved. The directive gives general policy guidance on the initiation, duration, and responsibility for configuration management. It also calls for joint configuration management when more than one DOD component (military service or agency) is involved in the acquisition, modification, or support of a product.

Two baselines are specified: a functional baseline to be established concurrently with approval to initiate engineering development; and a product baseline to be established immediately upon completion of audits of functional and physical configuration of each CI.

Concurrent with the release of DODD 5010.19 on August 6, 1968, DOD Instruction 5010.21, "Configuration Management Implementation Guidance," was approved. This document provides guidance for the implementation of DODD 5010.19. It expands the policy and provides some firm guidelines for implementation. This document adds an optional "allocated" baseline between the functional and product baseline established by the DOD directive. The allocated baseline is to be used to assign a complex configuration item's functional characteristics to its major components.

At the time this book is being published, the directive and instruction documents for configuration management are being reviewed by DOD, along with other policy areas dealing with major system acquisition, for possible restructuring. Also, the Air Force has proposed an additional standard: MIL-STD-483 (USAF), "Configuration Management Practices," to supersede AFSCM 375-1. This current activity is discussed in greater detail in Chapter 22, "The Contemporary Scene."

1.9 FUTURE OF CONFIGURATION MANAGEMENT

The continued application of formalized configuration management techniques for government-purchased products is a certainty because no other system is available for guaranteeing that detailed contract and product requirements are met. However, for private industry not connected with government projects, the future is less predictable. Nevertheless, it is apparent that the need is great if one is sensitive to the present sorry state of consumer product construction, reliability, and repair. Failure of industry to correct deficiencies, such as safety hazards and poor servicing of failed products, may lead to a consumer demand for more government controls. This demand could result within the next several years as a consequence of troublesome products, high prices, and a growing number of literate and aggressive consumers who insist on higher standards from industry.

Americans are rapidly becoming more alert to safety and quality as attested to by the popularity of books on auto safety, air pollution, drugs, medical care, and the use of chemicals in foods and crops. They are also becoming less impressed with flashy looking cars as is reflected by the large sales of foreign cars that lack the usual esthetic appeal. Conscientious and forward-looking industry leaders, who will but listen to the consumers' message voiced by a small number of articulate individuals, will begin a program to improve their products through the application of formalized configuration management techniques.

In summary, configuration management will continue to be a major factor in the design, development, and manufacture of products. The extent of the effort required will depend on the application of the product and customer requirements. It is expected that areas of industry not currently using formalized configuration management techniques will begin to apply these techniques on a growing scale to satisfy specific objectives, such as uniform performance and reliability of commercial test equipment or consumer products; for example, automobiles, appliances, and clothing. With this growth, designers and engineers of all types will require a sound knowledge of configuration management philosophy and methodology. The basic principles described in this book can be applied to all areas of industry. No matter how small a portion of this book is applicable, the detailed system and methods discussed should be modified to suit the specific technical and financial requirements of the user.

1.10   PLAN OF BOOK

This book is divided into four parts: Part I covers the details of configuration control; Part II discusses the configuration identification requirements and techniques; Part III describes the configuration accounting methods and documentation; Part IV discusses operational computer software controls, the product cycle, and the contemporary status of configuration management. As mentioned, a glossary and list of acronyms are included at the end of this book and may be referred to whenever unfamiliar terms appear in the text. A bibliography of government and industry publications on configuration management and related activities is also included.

PART I CONFIGURATION CONTROL

Chapter 2

ORGANIZING FOR CONFIGURATION CONTROL

As stated in Chapter 1, configuration control involves the "before-the-fact analysis of change requirements and feasibility of approach alternatives; impact on interfaces, resources, and schedules; commitment to action; and the assurance of controlled release of [configuration data at] all appropriate baseline points in the program cycle. Each change is in essence a new contract concerned not only with the projection of effort, but also with the effect on completed and in-process work."[1]

Before-the-fact analysis requires familiarity with project objectives, historical and planned configurations, and the change sensitivity of project elements. These data are available in the as-built configuration records and the design source documentation accumulated in developing the project baselines discussed in Chapter 1. Although this is an oversimplification, it should be pointed out that in order to control changes, it is necessary to know what the configurations are, which of these must be changed, how, when, where, and by whom.

2.1 CONFIGURATION CONTROL POLICIES AND OBJECTIVES

In responding to configuration control requirements, no really new or exotic disciplines need be imposed; instead, existing practices should be extended and systemized within the following limiting policies and objectives:

1. Only changes to fulfill technical requirements, to improve product performance, to reduce costs, or to provide advantages to the customer shall be proposed and implemented.

[1] Irving Mayer, Engineering Chief Minuteman Systems Division, Autonetics, NAA, *IEEE Transactions on Aerospace—Support Conference Procedures.*

2. All engineering and configuration data affected by formally approved changes shall be revised as necessary to describe the change.

3. Formally approved changes shall result in appropriate corresponding changes to all related support elements—manuals, handbooks, support equipment, etc.—to the extent authorized by the contract.

4. All actions related to the submission, analysis, approval, and implementation of changes shall be adequately documented.

5. Approved changes to engineering configuration data shall be properly implemented to assure that the cost of the change is minimized and that the configuration item will be produced as specified.

6. Retrofit engineering changes shall be recommended for incorporation only when mandatory for functional reasons or when retrofitting is to the buyer's advantage.

7. Configuration management disciplines shall be established and maintained on all subcontracted systems or major components designed by subcontractors.

2.2 ORGANIZATION

To implement and achieve the forgoing policies and objectives, configuration management requires a special organization that reports and is directly responsible to a manager who is independent of the engineering, testing, manufacturing, and quality assurance departments. This condition is necessary to avoid dilution of controls, which would normally occur because of the conflicting interests and objectives of these departments. In many large companies, a pool of configuration managers is available for assignment to specific projects. For smaller companies, the position of configuration manager is a staff function reporting directly to a corporate office or division manager.

In addition to a "separation of authority," adequate facilities and support personnel are necessary. Facilities include a data control center for distributing, maintaining, and storing records, drawings, specifications, and standards and for preventing unauthorized persons from obtaining and altering original vellums[2] without official permission to make changes from the project and configuration managers. Computer facilities may be needed on large projects to provide automated processing of configuration data and printout of tabulated reports. Support personnel include data control clerks, data analysts, configuration administrators and planners, reproduction operators, and secretarial-clerk support. A proved set of procedures for change control processing must be available, understood, and followed.

[2] Translucent reproducible sheet that contains original data.

Also, depending on the project scope, the configuration manager requires the support of a permanently established committee, a configuration control board (CCB), which is the final authority within the project on proposed major changes.

Again it should be noted that the configuration manager's position within the project management structure generally depends upon the size of the project. For small projects, the project manager is usually his own configuration manager. For larger projects, a separate configuration management office (CMO), under the direction of a separate configuration manager, is generally established.

The structure of configuration control boards is also variable. The typical organization includes representatives of engineering, production control, manufacturing engineering, purchasing, contracts, quality assurance, reliability, maintainability, and document/data control and is under the chairmanship of the configuration manager or project manager. It is highly important that board members are truly representative of their home departments and that they have authority to commit those departments. The board may also be chaired by the assistant project manager for systems engineering, product integrity, design and development, or by the project manager himself. A CCB of such structure is dedicated to *one project*.

However, there is a growing trend today to install a central configuration control board which assumes the responsibilities over several projects when these projects are not complex in structure. This type of grouping requires somewhat careful consideration.

In these days of structured (matrix) functional organization with superimposed projects and personnel having possibly two or three reporting relationships the board managers and representatives need to be adroit indeed. The central configuration control board would be constantly working with functional organizations having varied interrelationships and would be called upon to interweave differing project requirements. This constant association, however, could be an asset itself inasmuch as there would be a certain amount of repeatability: Insights gained during one foray would pay off for the next. It would also be wise that a consolidated CCB have as projects those that have the same disciplines and product line characteristics involved; for example, several projects involving instrumentation. In this respect subcontractors selling product line items to various associate contractors often use a single company CCB.

Although configuration control boards may differ in size, participation, and level of activity, they have in common the responsibility for the direction of complete change evaluation, change planning, and change incorporation.

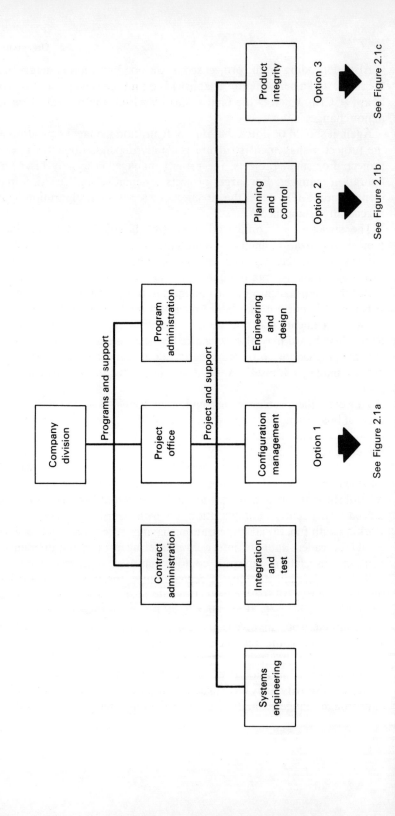

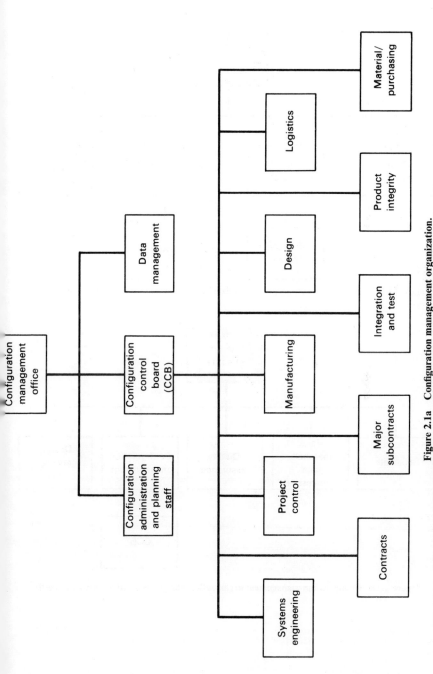

Figure 2.1a   Configuration management organization.

The same basic configuration management organization forms Options 2 and 3. The following figures show the relationship of the organization within planning and control and product integrity.

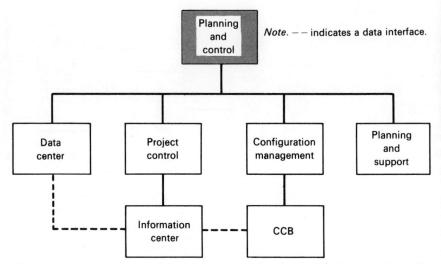

Figure 2.1b   Configuration management organization when placed within planning and control.

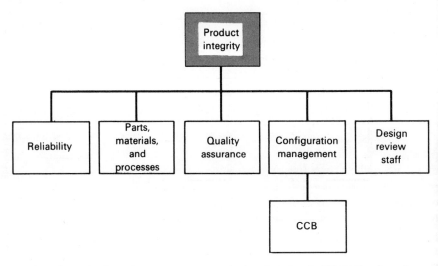

Figure 2.1c   Configuration management organization when placed within product integrity.

2.3 TYPICAL ORGANIZATIONS

Three options in organizing for change control are shown in Figure 2.1. The primary difference is the management level at which the configuration manager functions. Although we may be biased and therefore choose Option 1 with the configuration manager reporting directly to the project manager, the general rule in industry has been that he will report *through* either the planning and control function or the product integrity (product effectiveness, product assurance, etc.) activity.

In addition, there are many approaches to organizing the configuration manager's own staff. We choose to group his responsibilities into three categories, each requiring a staff leader: configuration administration and planning, configuration control board (CCB), and data management. Small projects, of course, would consolidate the responsibilities described in this example and delegate them to only one or two people in support of the configuration manager.

2.4 THE PROJECT MANAGER'S ROLE

Project managers are totally responsible for all phases of contract performance. One of the very significant phases is the configuration management of the project's deliverable end items. To discharge this responsibility, the project manager relies on personnel assigned to his direct supervision in the project office and on personnel appointed by the various functional organizations to serve the project. The project manager establishes the configuration management techniques necessary for his project to meet customer and company requirements within the resources available to him. These guidelines are published as a "configuration management plan" and form a part of his overall project plan.

In some cases the configuration management plan is available from other projects or is a standard company document that applies to all projects unless it conflicts with the project manager's particular contract requirements. The plan is normally prepared by the configuration manger, and it defines the philosophy, objectives, organization, approach, policies, and forms that will be followed during the project. The plan is based on the previously mentioned standard, if it exists, which is modified or adopted to meet special project manager or customer requirements. To optimize project configuration management performance, existing company procedures should be followed whenever possible.

The project manager appoints the configuration manager and establishes the configuration control board (CCB) at the inception of his project. It is likely that the configuration manager has already participated in the preproposal effort and earlier definition phase. Since the chairman of the CCB is the decision-maker on proposed changes to the product(s), the soundness of his judgment profoundly influences the success of the project. The project manager should therefore either personally act as CCB chairman or delegate this authority to the assistant project manager for systems engineering, as an example. He could also appoint the configuration manager if he is reporting directly to the project manager (Option 1, Figure 2.1a) or if the configuration manager is a senior engineer as well as an administrator.

Once the configuration management system is initiated, the project manager gradually reduces his participation in day-to-day configuration management activities and the configuration manager takes on full responsibility for the details of controlling, identifying, and recording changes. When critical technical, schedule slippage, major fiscal, or emergency situations arise, however, the project manager may again become fully involved in order to define, control, and authorize special procedures to avoid delivery delays or to sustain required performance and design requirements within the framework of the CCB.

It must be pointed out, however, that the majority of changes *do not* require processing through the CCB. They do, nevertheless, require a collective evaluation for impact and the formulation of effective planning for their incorporation. These changes are by far the configuration manager's greatest challenge.

2.5   CONFIGURATION MANAGER

We have earlier assigned, for the sake of reference, three categories of responsibilities to the configuration manager: configuration administration and planning, configuration control board operations, and data management. It is tempting to make the more simple suggestion that three staff group leaders be assigned to the configuration manager, each respectively responsible for identification, control, and accounting (and many projects do just that); however, with respect to the case example chosen, the latter familiar terms appear too arbitrary. You will find as many variations as there are contractors, each adding particular checks and balances in support of what works for his company. Let us reestablish some of the configuration manager's overall principal objectives before we further explore the three subsets of responsibility suggested.

Design Authority Documentation System.   An early objective of the configuration manager is a formal system of design documentation for recording the technical requirements of each end item of hardware beginning with the initial concept of the functional performance requirements, breadboard design, and preliminary drawing trees, and progressing in orderly evolution to the creation and delivery of an item in its installed operating environment.

Release and Change Control Procedures. High on the configuration manager's list of objectives is the comfort of a well-defined project management system for the release of design authority documentation[3] (drawings, parts lists, specifications, test procedures, etc.) and change control procedures, which include the following:

1. An orderly identification and scheduling system for initial release of design data and expeditious accomplishment of changes once the drawing or specification is released. Major objectives are the following:

   a. Integrate the changes among all affected operations.
   b. Establish a change introduction point and disposition for existing materials.
   c. Assure changes as defined are correct, clear, complete, and understood by responsible personnel.
   d. Assure proposed changes are in accordance with overall project direction and authority.

2. A procedure for nonengineering groups to request special changes.
3. A procedure for subcontractor change control.
4. A central location (integrated record system) where the status of existing hardware is known; for example,

   a. Quantities of hardware in existence.
   b. Stage of completion of hardware and its location.
   c. Where hardware is used.

End Product Identification. Consistent and systematic methods must be applied to the identification of baselines, physical hardware, parts, and subassemblies, as well as to the documents and the data changes that both authorize and support the hardware and define precise baselines.

Integrated Record System. Attention must be given to an integrated record system which will contain all pertinent configuration identification data generated during the design, the manufacture and inspection, and the acceptance test period.

1. Design authority records which establish the planned configuration.
2. Engineering and manufacturing summaries which determine the engineering configuration and assure that authorized changes have been incorporated in the *actual* manufactured article. These also record the location of critical[4] or time-sensitive components[5] in each article in order to relate failure data and maintenance data to the component's manufacturing, test, and operating history.

---

[3] Also referred to as source documents.
[4] An item essential to the operation of the CI.
[5] Components that degenerate with time (finite shelf life).

2.6 CONFIGURATION ADMINISTRATION AND PLANNING STAFF

The configuration administration and planning staff, which supports the configuration manager, performs the housekeeping function of the CCB and the implementation planning for all changes. Its administrative responsibilities include the following:

1. Receive, number, log in, and process to completion all engineering change requests initiated on the project.

2. Prepare agendas, arrange CCB meetings, and act as secretary to the chairman in preparing minutes or other directives resulting from CCB activities.

3. Where it is a contractual requirement, process all engineering orders[6] to the local customer representative for his review. Process the complete technical data package for all engineering proposals to the point of turning them over to the contracts department for ECP submittal to the customer.

4. Receive notification of customer approval or disapproval of engineering proposals from the contracts department and take all necessary action to either release definitive engineering data as proposed or to stop all further activity on disapproved changes.

The prerelease planning responsibilities of the configuration administration and planning staff are primarily directed at the majority of project changes, discussed earlier, which do not process through the CCB. For this purpose a small three-man board [change planning group—(CPG)] may be established at the option of the configuration manager. The CPG will serve as a detailed prerelease implementation planning group for determining information on the effects of changes to design manhours, test requirements, manufacturing costs and schedules, and the complete effect on logistics support requirements. The determination of the planned incorporation point or effectivity of a change and its attendant parts and materials disposition instructions is a primary CPG responsibility.

The CPG will also function in direct support to CCB agenda items. The CPG is composed of the configuration administration and planning representative, a manufacturing product-line planner appropriate to the configuration item under evaluation, and the engineering design representative for the CI. The prerelease planning of configuration changes is, of course, a mandatory project activity; however, the scope of this activity is a function of product complexity and of the volume of design change activity. Small projects and projects well into the production and operations phases may elect to accomplish the overall functions of the configuration administration and planning staff with a single individual supporting the configuration management office.

---

[6] A document that describes an engineering change to a released drawing, parts list, or wire list.

The postrelease responsibilities of the configuration administration and planning function are directed at the verification and approval of all engineering orders, specification revisions, and related change documents with respect to the effectivity instructions determined during CCB or CPG prerelease planning.

2.7 CONFIGURATION CONTROL BOARD

The CCB is not a voting board; decisions are the sole responsibility of its chairman, although data and arguments regarding proposed changes are presented by representative members from all major functions. The board is a permanently established committee (see Figure 2.2) which is the final authority within the project on proposed major (Class I and complex Class II)[7] changes. The principal function of each board member is to

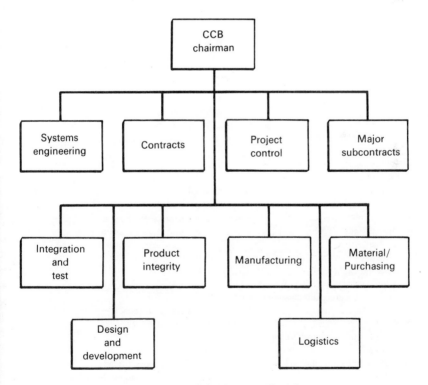

Figure 2.2   Configuration control board.

---

[7] A Class I designation is used for a change that is beyond the scope of the contract. A Class II designation is used for a change that is within the contract scope.

verify the following:

1. The change is necessary.
2. The method of implementation is feasible.
3. The member can meet schedule and cost requirements.

If the existing design deficiency prevents the equipment from working properly, changes must be made to correct the deficiency, and the project is usually responsible for the added cost. However, if the proposed change is a design improvement, and therefore out of the scope of the contract, an official engineering change proposal request (ECP) is submitted to the customer to obtain his approval for the added cost and schedule delays. Table I given below is a checklist for ensuring that the board members have considered all aspects of the proposed change and its effects on other project areas.

TABLE I
Configuration Control Board Checklist

1. How does this change affect total CI performance?
2. Are other projects or CI's in the system affected by the change?
3. How does the change affect the reliability of the CI?
4. Can the change be implemented by manufacturing as proposed?
5. Is special tooling required and how much will it cost? How long will it take to make this tooling?
6. Will previously delivered CI's have to be returned for rework or will they be field (in-service) retrofitted?
7. How will the CI change affect the delivery schedule?
8. What affect will the change have on the total project cost?
9. What drawings, specifications, manuals, and procedures will have to be changed?
10. How are the CI weight, size, balance, stability, and power consumption affected by the change?
11. Is the CI or operator safety affected?
12. Does the change affect spare parts and assemblies?
13. Is the service life of the CI affected?
14. Are repair and maintenance made more difficult by the change?
15. Will the magnetic and radio interference characteristics of the CI be changed?
16. Will the mechanical and electrical installation of the CI be affected?
17. Will the qualification CI have to be modified?
18. Will parts procurement present any problems?
19. Will completed assemblies without the changes have to be scrapped?
20. What is the effectivity[a] of change incorporation? Which serial numbered assemblies must have the changes made to them?
21. Will additional tests be required? What is their schedule and cost impact?

[a] The approved use of a change in specific CI serial numbers.

Class I Engineering Change Criteria

The primary attention of the CCB is directed at the processing, planning, and submittal of Class I changes. Class II changes are usually processed by the CPG. Class I engineering changes submitted for customer approval are limited to those which are necessary or offer significant benefits to the customer. Such changes are those required to accomplish the following:

1. Correct deficiencies.
2. Make a significant effectiveness change in operational or logistics support requirements.
3. Effect substantial life-cycle cost saving.
4. Prevent slippage in an approved production schedule.

The preceding criteria provide general guidelines for identifying Class I changes. However, specific guidelines are required for identifying an engineering change as a Class I change. These guidelines[8] are listed below. (Any changes not affecting items 1 through 5 are identified as Class II changes.)

1. The functional or allocated configuration identification (performance specifications, etc.).
2. The product configuration identification (design specifications, drawings, etc.).
3. Technical requirements below contained in the product configuration identification, including reference data.

   a. Performance outside stated tolerance.
   b. Reliability, maintainability, or survivability outside stated tolerance.
   c. Weight, balance, moment-of-inertia outside limits.
   d. Interface characteristics altered.

4. Nontechnical contractual provisions.

   a. Fee.
   b. Incentives.
   c. Cost.
   d. Schedules.
   e. Guarantees or deliveries.

5. Other factors.

   a. Government furnished equipment (GFE).
   b. Safety.
   c. Electromagnetic characteristics.
   d. Operational, test, or maintenance computer programs.

---

[8] MIL-STD-480, "Military Standard, Configuration Control—Engineering Changes, Deviations, and Waivers."

e. Compatibility with support equipment, trainers, or training devices/ equipment.

f. Configuration to the extent that retrofit action would be taken.

g. Delivered operation and maintenance manuals for which adequate change/revision funding is not on existing contracts.

h. Preset adjustments or schedules affecting operating limits or performance to such extent as to require assignment of a new identification number.

i. Interchangeability, substitutability or replaceability, as applied to configuration items (CI's) and to all subassemblies and parts of repairable CI's, excluding the pieces and parts of nonrepairable subassemblies.

j. Sources of CI's or repairable items at any level defined by source control drawings.

2.8 DATA MANAGEMENT GROUP

The data management group duties are the management of engineering documentation, including its identification, recording, maintenance, care, storage, retrieval, and distribution to project members and to the customer. The group must function as a counterpart to the government data agency administering the contract. Data to the customer consist of a customer data package. In some contracts the terms and contents of the package are described by a data checklist.[9] The group is also responsible for the important task of designing and installing the integrated record system discussed earlier. To repeat, the basic aims of the integrated record system are as follows:

1. To establish current and rapid retrieval records of the planned or as-approved design configuration described by design authority documents; that is, specifications, drawings, test procedures, engineering orders.

2. To ensure that as-built configuration records are compiled and maintained to facilitate certification of change incorporations and identification of waivers, test deviations, parts substitution, etc. (configuration verification). These records are based on information contained in the engineering drawings, engineering orders, and specifications used in the production processes and all associated manufacturing, inspection, and test records.

3. To assure the availability of records on the location of critical or time-sensitive components in each article in order to relate failure data and maintenance data to the component's manufacturing, test, and operating history.

[9] AFSCM 310-1 and AFCM 310-1, "Management of Contractor Data and Reports."

If special customer requirements on parts and material traceability due to the hi-reliability nature of the product are imposed, the above capability would extend forward into all levels of vendor processing, that is, purchase orders, lots, mill runs, batches.

The efforts associated with the requirements for configuration verification and parts or materials traceability should make full utilization of existing manufacturing, inspection, and test documentation. Because of the "tyranny of numbers" inherent in realizing the objective of rapid retrieval records, large projects will require the application of automatic data processing techniques. A discussion of both the manual and automated aspects of an integrated record system appears in Chapter 21.

A number of the detailed responsibilities of the group relative to engineering data management are described below.

Determination of Drawing Levels.   Drawing levels or grades to be applied to engineering model, prototype, qualification model, and operational CI drawing packages are determined by the data management group. These levels indicate the drafting quality and standards that will be followed during preparation of the drawings; for example, an engineering model drawing can be made to less stringent standards than a production drawing. Revision mechanisms for each level are also selected. Since these requirements are very important to design engineering, all plans for the drawing release program are developed in close coordination with the configuration manager, design supervisor, and project manager.

Issuance of Identification Numbers.   Identification numbers are issued for hardware and software under a standard controlled company system. Log books are maintained for numbers issued to products, drawings, test procedures, manuals, reports, parts lists, specification control drawings, data lists, specifications, engineering orders, engineering change proposals, and so on.

Identification of Baselines.   Baselines are identified by publishing baseline reports, as defined in the configuration management plan. These reports describe the CI configuration at each baseline by identifying each drawing, revision letter, and part number used to describe each item in the product/CI.

Compilation and Issuance of Configuration Reports.   Configuration reports, such as the baseline report above, are compiled and issued by the data management group. The reports are prepared to the format and schedule specified by the contract. These reports identify all approved changes that are in process or completed for each CI and describe the current configuration of each CI by listing the part numbers and drawing revision letters for each item that goes into the CI.

Maintenance of Records.   Records are maintained for each engineering document release and issuance. In addition, records are kept of each change, its effectivity, and verification of compliance with effectivity instructions.

Development and Processing of Special Documents.   Special parts lists, data lists, and configuration record summaries are prepared and processed by the data management group as required by the configuration management plan. The fulfillment of this responsibility ensures that data list information is correctly prepared for mechanized publication of data, or, when the mechanized system is not used for the project, that the required lists are prepared manually. Typical of such lists are the following:

Product allocation list, which identifies the planned systems and/or geographic location of each CI and the quantity requirements at each location or in each system.

CI parts list, an indentured or alphanumeric listing of parts, materials, and assemblies forming the CI.

Project data list, which identifies all technical and management document requirements by quantities, scheduled need and delivery dates, and contractual task. Interrelates individual CI data lists.

Customer configuration status accounting records, which are periodic reports on the status of customer approved changes to in-process (production) and delivered CI's.

The data management group is usually supported by a centralized company data control department. If not, a support group must be formed to fulfill the following detailed responsibilities relative to the handling, care, reproduction, and distribution of engineering documentation:

1. Issue properly approved data upon authorization.
2. File or microfilm copies of documents and revisions or changes for historical records, distribution, and disaster file.
3. Issue and distribute approved changes and revisions to all holders of controlled copies of data.
4. Supply reference copies on a demand basis.
5. Maintain historical records of data status, usage, and distribution.
6. Store and safeguard originals of data.
7. Audit company data usage and controls.
8. Issue change numbers or letters for data.
9. Procure required government and industrial engineering publications applicable to the project.

Chapter 3

DESIGN AUTHORITY DOCUMENTATION

Seven kinds of design authority documentation provide source data for engineering, test, and support groups. These documents are:

1. Advance drawing breakdowns (list of drawings to be used for CI) and advance bill of materials.
2. Drawings.
3. Specifications.
4. Parts and data lists.
5. Test procedures.
6. Technical operating manuals.
7. Shipping and handling instructions.

These source data are constantly changed by refinements fed back to design groups pertaining to mockup[1] and breadboard development information, prototype design improvements, test responses, and major changes in manufacturing sequence. The following sections describe each type of design authority documentation.

3.1 ADVANCE ENGINEERING RELEASES

Engineering information is usually on hand for scheduling and planning many procurement, production, test, and support functions well before detailed drawings exist. This "advance release" point for each CI or product should be established as a scheduled point within the project plan. The scheduled point may vary in time for each CI depending on the commonality of the item to previous designs on the one hand, and its state-of-the-art objectives on the other. The methods of making the advance information

[1] A simulated model of the CI's mechanical or thermal configuration.

43

Symbols:
—— Planned Effort ── Scheduled in Check ──△ Scheduled Release ──⟁ Previous Scheduled Release
━━ Actual Effort ── Actually in Check ──▲ Actual Release

Item No.	Drawing Number	Size	Assembly Level	Drawing Title	In Work	In Check	Released	Jan	Feb	Mar	Apr	May	Jun	Jul	Aug	Sep	Oct	Nov	Dec
1																			
2																			
3																			
4																			
5																			
6																			
7																			
8																			
9																			
10																			
11																			
12																			
13																			
14																			
15																			
16																			
17																			

Remarks:

Approvals

Product Engineer _____
Drafting _____
Checking _____
Project Office _____

Application

CI _____
Used On: _____
Sheet ___ Of ___
Revision Date _____

(ABC) Corporation   Code Ident. 09876

Project: _____   Rev. _____

IDL (Configuration Item Drawing Number)

IDL
Sheet ___ of ___
Rev. ___

available are as follows:

Design Layouts

As the new design is developed, design layouts are made to establish the basic lines and structure of major assemblies, mechanisms, functional systems, and installations. These are conceptual drawings but are usually in sufficient scale and detail to permit detailed drawing requirements to be known.

Indentured Drawing Lists and Schedules

An indentured drawing list is prepared for each CI from the design layouts. Through it can be established the indentured relationships of assemblies and the schedule timing of drawing releases for the CI. See Figure 3.1. Indenturing is the process of starting with the CI and systematically identifying and listing the items that comprise the CI by showing their subordinate positions with respect to the CI and its assemblies. These items are often listed indented from the left-hand side of the page, the amount of indenting depending on whether the item is directly assembled into the CI or into a lower level component which is then assembled into the CI. For example:

Indenture Level							
1	2	3	4	5	6	Drawing No.	Title
X						123450	Top assembly or CI
	X					123500	Subassembly no. 1; goes into CI
		X				123505	Component no. 1 of Subassembly 1
		X				123508	Component no. 2 of Subassembly 1
			X			123600	Module that goes into Component 2
	X					123700	Subassembly no. 2; goes into CI

Thus, as items are categorized into lower levels of assemblies, the indenture level or number increases. An arbitrary indenture level of 6 is given here, but it could have been more, depending on the complexity of the CI and other configuration requirements.

The change mechanism for the indentured drawing list should be as flexible as possible, in that its early availability and currency is of major importance to the configuration management office, manufacturing planning, tooling design, and quality and test engineering. Drawing release schedules entered on the indentured drawing list are developed in conjunction with the project master schedule. These drawing schedules should be negotiated with manufacturing to determine the sequence in which drawings must be available for production planning. Drawing release schedules are usually developed

Item No.	Part Number	Part Name Description	Specification	Manufacturer	Source Document Number	Rev.	Hardware Quantity	Support Quantity	Total Quantity	Estimated Cost	Need Date	Lead-Time (weeks)	Remarks
1													
2													
3													
4													
5													
6													
7													
8													
9													
10													
11													
12													
13													
14													
15													
16													
17													
18													
19													
20													

Remarks:

Approvals

Product Engineer _____
Material Control _____
Other _____
Project Office _____

Application

(Configuration Item) _____
Used On: _____
Sheet _____ Of _____
Revision Date _____

(ABC) Corporation

Project _____

ABM

Code Iden. 09876

Rev. _____

(Configuration Item)
(Drawing Number)

ABM
Sheet _____ of _____
Rev. _____

Figure 3.2 Advance bill of materials.

for (a) the engineering model release; (b) the qualification model release; and (c) the prototype and operational model release. In the latter two cases the prototype/operational drawing complement is usually an update of the qualification model description.

Advance Bill of Materials

The advance bill of materials provides a means for procurement action considerably in advance of formal drawing release. Many purchased components require unique fabrication techniques by subcontractors, and advance material commitments are necessary to allow sufficient design and manufacturing leadtime. The sample advance bill of materials form illustrated in Figure 3.2 is used to list the "full complement" of components making up the CI known to the design engineer at a point in time. The list is updated as often as possible, by "red-line" [2] methods if necessary, to keep it as current as possible. The combination of lists from other "related" CI's will provide the advantage of integrated buys within, at the minimum, individual product lines and subsystems. Responsibility for the decision to order advance items with respect to accepting cost, subsequent change, and obsolescence risks is shared by the project manager, design engineering, and procurement.

Equipment Allocation Lists

Equipment allocation lists prepared by the configuration management office serve to establish advance allocation commitments of configuration items to the engineering, qualification, prototype, and operational configurations, and attendant spares and special test articles. The equipment allocation list is interrelated with the work breakdown structure [3] and specification tree [4] for the project.

3.2 DRAWINGS

Engineering drawings are pictorial, tabular, and narrative descriptions of a portion of the CI being manufactured for the customer. A drawing usually consists of a title block, list of material, usage or application data, change history (revision block), general notes, and the drawing itself with appropriate dimensions. See Figure 3.3.

---

[2] Changes are made on the list in red pencil rather than by retyping.

[3] Shows tasks, services, software, and hardware required to complete project. Refer to MIL-STD-881.

[4] Shows relationships of project specifications.

Revisions				
Zone	Letter	Description	Date	Approved
B4	A	$\frac{1}{16}$ R was $\frac{7}{32}$ R	3-6-70	J. Jones
C4	B	1.75 was 1.74	5-7-71	J. Jones

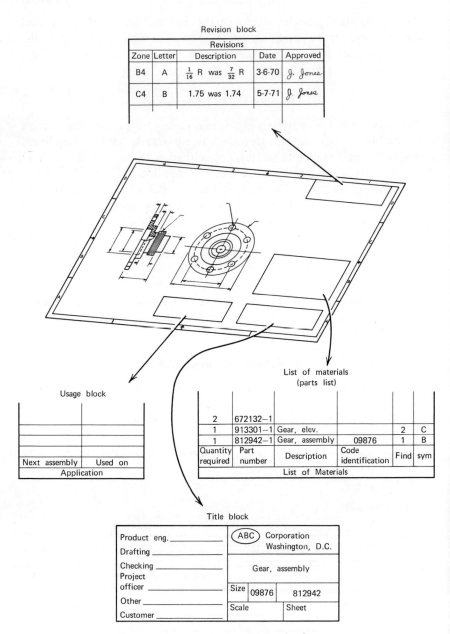

Usage block

Next assembly	Used on
Application	

List of materials
(parts list)

Quantity required	Part number	Description	Code identification	Find	sym
2	672132−1				
1	913301−1	Gear, elev.		2	C
1	812942−1	Gear, assembly	09876	1	B
List of Materials					

Title block

Product eng. _____	(ABC) Corporation Washington, D.C.	
Drafting _____		
Checking _____	Gear, assembly	
Project officer _____		
Other _____	Size 09876	812942
Customer _____	Scale	Sheet

Figure 3.3 Typical engineering drawing format.

Engineering drawings must be comprehensive, thorough, self-explanatory, and able to stand on their own without benefit of the originator's presence. During the life of the contract, orderly communications among management, engineering, production, quality, suppliers, and any other users can only be maintained through this graphic language of properly updated and complete drawings. Typical of the drawings found in the aerospace industry are the following:

1. *Assembly*, showing component parts suitably identified in their proper positions.

2. *Detail*, a single, self-contained part with design information.

3. *Installation*, a form of assembly drawing showing the attachment of subsystems, equipment, and hardware to the main frame of the system.

4. *Source (or specification) control*, establishing the approved source(s) of a procurable component or part on an assembly drawing that has been designed for a specific application. May include applicable configuration, design, and test specifications.

5. *Design layout*, establishing the basic lines of major assemblies, mechanisms, and installations. These are conceptual drawings of sufficient detail to define drawing requirements and item assembly relationships.

6. *Tabulated drawing*, showing a part that is to be fabricated or procured in various sizes. The part outline is shown in a single picture, with its variable dimensions designated by dash numbers. These numbers are shown in a table in which the actual dimensions are tabulated for each application.

7. *Tubing and cable*, depicting *routing*, components and their physical locations, arrangements, diameter, type and size of fittings, and materials.

8. *Wiring diagram*, an undimensioned drawing that shows the interconnection of components in a CI, usually electrical or hydraulic.

9. *Ground equipment*, drawings of support, test, and maintenance equipment making up an operational system.

10. *Packaging and crating*, auxiliary drawings showing provision for shipping.

11. *Schematic*, a diagram that shows the electrical connections and functions of a specific circuit by graphical symbols. Mechanical schematics show the operational sequence or the mechanical arrangement of a device.

12. *Top assembly*, the highest drawing describing the CI, containing three views of CI, a list of major assemblies, an identification plate, and hardware for attaching the CI to its higher level assembly.

3.3 TYPICAL DRAWING DESCRIPTION

The most common example of an engineering drawing is the detail assembly drawing as shown in Figure 3.4. This delineates detailed parts for which there are no other drawings and also delineates the combining of

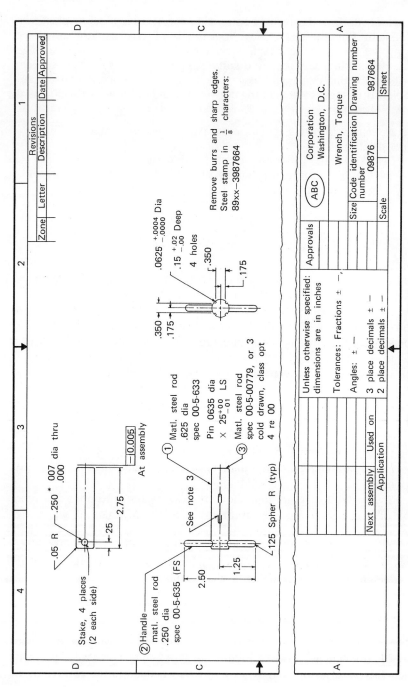

Figure 3.4 Detail assembly drawing.

these parts, along with others, into an assembly drawing, which is used to machine and fabricate the detailed parts and their parent assemblies.

Graphic Presentation of Design

The body of the drawing is the graphic representation of the design along with the required dimensional information and some unique identification of the components shown. The identification is usually the part number or a "find" number keyed to the parts list or list of material.

Title Block

The title block area is used to record information about the drawing number, drawing title, drawing sheet size (in the form of a letter code), scale of drawing, sheet number, manufacturer's code number (see Chapter 9), and authorizing signatures required before the drawing can be officially released.

Revision Block

The revision block area is used to record or refer to the written description of changes made to the drawing after its original issue. Each change is identified by a change letter to distinguish between revised issues of the drawing.

Usage Block

The usage block is sometimes referred to as the next assembly block or the applications block. It is used to indicate the usages or the application of the item on the drawing. The usage data consist of the next higher assembly and the configuration item and model that will use the item. This information is listed against each configuration of the item established by the drawing. The usage block may also include the listing of the CI serial numbers to which the application of the items and its various configurations is directed.

The relationship of drawing usage data to the list of materials (parts list) for the next assembly is discussed in Chapter 8.

List of Material or Parts List Block

The list of material block is used to indicate the composition of assemblies as well as the engineering material description of the detailed parts created on the drawing. It consists of the parts list, which lists the assembly components, and an actual list of material as defined by engineering. The parts list identifies each component or item of the assembly by number and name and indicates the quantity of each required to make an assembly. A source[5] or related source control document is indicated for purchased parts. Also, for each detailed part created, it contains the material description, final cut size, and specification reference.

3.4 DRAWING GRADES AND SIZES

MIL-D-1000 and MIL-STD-100[6] establish three grades or levels of drawing preparation for government contract work. Irrespective of the government requirements, the application of varying levels of drawing format standards within a company is practical and useful. The representative levels of MIL-D-1000 are described in the following paragraphs.

Form 1 Drawings

The form 1 drawing is prepared to the highest standards required by the contract (MIL-STD-100) or company and is used for the production and logistics support of the product. The form 1 drawing requires a formal drawing review and the signatures of the following project functions:

1. Engineering.
2. Project office.
3. Manufacturing.
4. Design and drafting.
5. Quality assurance/reliability.
6. Specialty areas, such as stress, weight, etc.
7. Checking.

The signatures required can vary to suit the company, project manager, and contracted requirements. Therefore the above list can be either shortened or lengthened to meet specific needs. However, drawing approval criteria are usually strictly defined and enforced as a company standard.

Form 2 Drawings

Form 2 drawings are prepared to industry standards. Form 2 drawings are usually used for development hardware (qualification or engineering model) or other nonflight equipment. Form 2 can also be used to order long leadtime items where the drawing or design is not complete but is

[5] Vendor, supplier, or subcontractor.

[6] MIL-D-1000, "Drawings, Engineering and Associated Lists." MIL-STD-100, "Engineering Drawing Practices."

adequate for beginning the procurement cycle. Release may require signatures of the project manager, design engineer, draftsman, and checker.

Form 3 Drawings

A Form 3 drawing is prepared to standard format but is an informal document which may include an engineering sketch. A drawing review is not required and release can be made on the signature of the design engineer. Form 3 drawings are used for research, nondeliverable, and advance technology products.

Drawing Sizes

Drawing sizes are identified by letter as shown below:

Drawing Size	Dimensions (Inches)
A	$8\frac{1}{2}$ by 11
B	11 by 17
C	17 by 22
D	22 by 34
E	34 by 44
F	28 by 40
G	11 by 42 to 144 } Roll size
H	28 by 48 to 144 } Roll size
J	34 by 48 to 144 } Roll size
K	40 by 48 to 144 } Roll size

3.5 SPECIFICATIONS

Chapter 1 introduced the concept of the progressive definitization of the product through the evolution of engineering data. Strongly associated with the baseline concept are the development and application of gradually "hardened" specifications. This is accomplished by the release and control of a system or total product specification early in the project, supported by subsystem, interface, environmental, and other support specifications—and most important, specifications for the individual major items (configuration items) comprising the system or product. The first release of the CI specification is performance in nature; that is, it describes the performance characteristics and development test criteria which must evolve into a detailed design for production. The second release is a design specification which fully establishes the production configuration and acceptance test program and is supported by production drawings, process and material specifications, production test procedures, etc. (See Figure 3.5.)

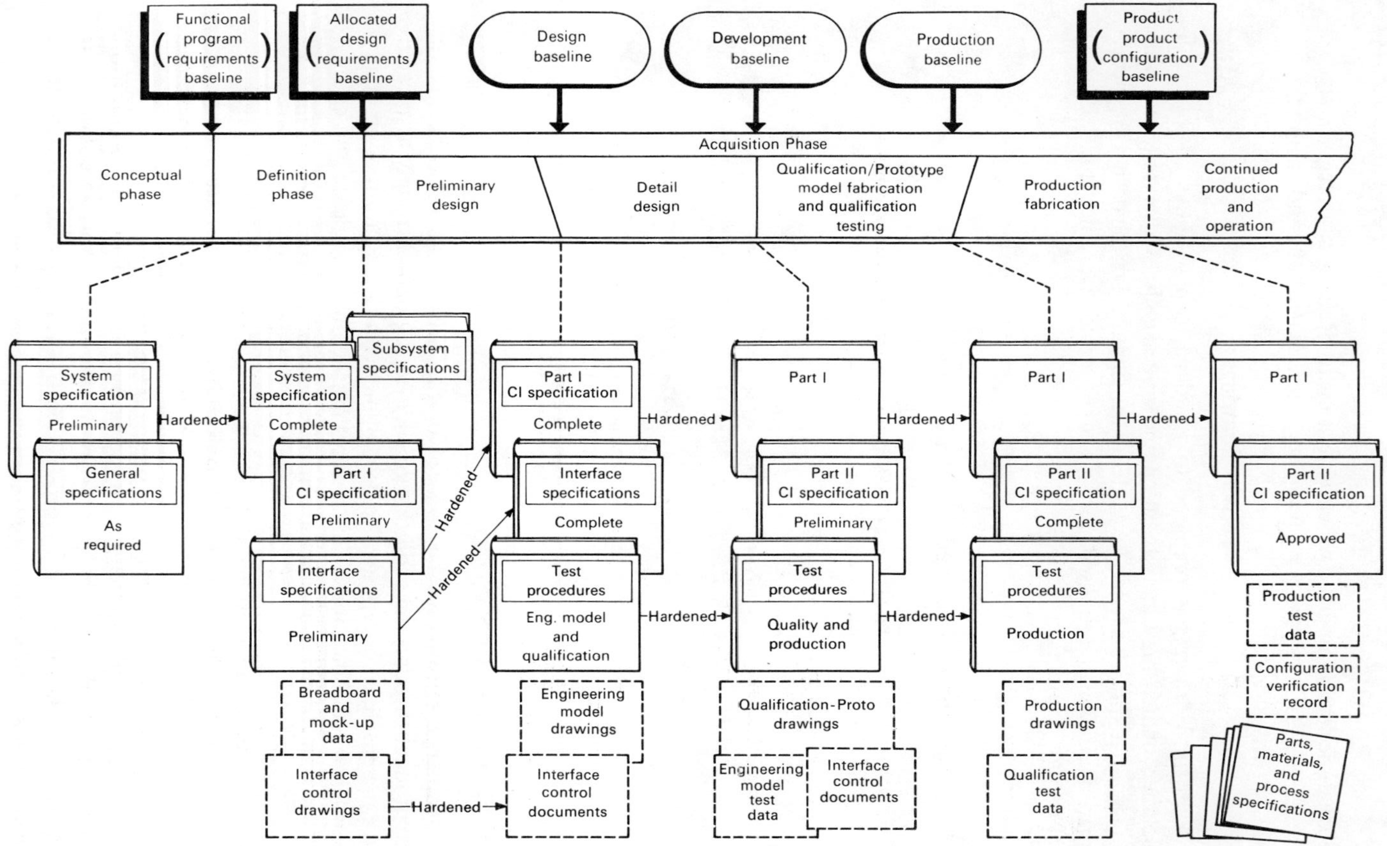

Figure 3.5 Specification development progression.

Before the 1960s the initiation of these two versions of a specification for a "black box" [7] CI (performance vs. design) was in response to two different military specifications; one dealing with "development goal" or performance specifications and the other establishing the requirements for a separate and distinct design specification. Both were delivered and maintained but, in large, each version was out of step with the other. Thus they were rarely used as day-to-day contractual instruments of reference for firm and current seller-buyer technical agreements.

This lack of continuity between performance specifications originated early in project activities, and the product (design) specification which appeared during the qualification test period, near the commitment to production, was attacked by the Air Force configuration management program established by AFSCM 375-1. The Air Force authored the Part I (performance) and Part II (product) concepts of correlated and transitional specification development as an integral part of the baseline management process. From the earliest release of the system specifications through the development of the Part I and Part II CI specifications into "one cover," the specification requirements for the CI were required to be progressively refined, managed, and applied as contractual points of agreements. Therefore it was possible to apply engineering change proposals and their attendant specification change notices (SCN) to the firm's specification description (whether only performance objectives or design absolutes) existing at any point in time. See Tables II and III for the typical contents of each specification.

The recent release of MIL-STD-490 has formalized the application of specifications into specific baseline points. The specification categories of the standard also closely approximate the Air Force specification categories. The typical types of government and industry specifications are illustrated in Figure 3.6 and discussed below. Chapter 10 presents specification identification and release requirements.

Contract (Government) Specifications

1. *System Specification*

The system specification establishes functional and performance and design criteria for the design, development, testing, and production of a complete system. The system specification allocates the system into functional entities identified as configuration items.

---

[7] A black box is a product element that identifies a functional requirement but does not reveal anything about its contents or construction.

TABLE II
Performance Specification Outline—Part I

Section No.	Title
1.0	Scope
2.0	Applicable documents
3.0	Requirements
3.1	Performance
3.1.1	Operational characteristics (includes measurement range, linearity, gain, accuracy, etc.)
3.1.2	Operability
3.1.2.1	Reliability
3.1.2.2	Maintainability (repair and servicing requirements)
3.1.2.3	Useful life
3.1.2.4	Environmental requirements
3.1.2.5	Human factors requirements
3.1.2.6	Magnetic and electric field cleanliness[a]
3.1.2.7	Safety
3.2	CI definition
3.2.1	Interface requirements
3.2.2	Component identification
3.3	Design and construction
3.3.1	Materials
3.3.2	Standard and commercial parts
3.3.3	Moisture and fungus resistance
3.3.4	Corrosion of metal parts
3.3.5	Interchangeability and replaceability
3.3.6	Workmanship
3.3.7	Identification and marking
3.3.8	Storage
4.0	Quality assurance
4.1	Integrated project test requirements
4.1.1	Engineering test and evaluation
4.1.2	Qualification testing
4.1.3	Reliability testing
4.2	Integrated test program test requirements at launch site
5.0	Preparation for delivery
5.1	Packaging and shipping
5.2	Transportability
6.0	Notes
7.0	Appendices

[a] Equipment (product) does not emit magnetic and electric fields.

TABLE III
Product (Design) Specification Outline—Part II

Section No.	Title
1.0	Scope
2.0	Applicable documents
3.0	Requirements
3.1	Performance (same as 3.1 in Table II)
3.2	Equipment configuration
3.2.1	Manufacturing drawings
3.2.2	Customer furnished property list[a]
3.2.3	Manufacturing standards (quality requirements for soldering, welding, etc.)
4.0	Quality assurance
4.1	Equipment performance and configuration requirements and quality assurance verification cross reference index
4.2	Test verifications
4.2.1	Engineering testing and verification
4.2.2	Preliminary qualification test
4.2.3	Formal qualification test
4.2.4	Reliability tests and analyses
5.0	Preparation for delivery
5.1	Preservation and packaging requirements
5.2	Packing and shipping container requirements
5.3	Shipment instructions
6.0	Notes
7.0	Appendices

[a] Items supplied by the buyer for inclusion in the CI.

2. Subsystem Specification

The subsystem specification establishes functional and performance requirements and design criteria for any major combination of equipment that performs a major function and is essential to the completeness of a system. Subsystem specifications are not normally employed in government contracts, but may be negotiated into being.

3. Equipment or CI Specifications

The equipment specification establishes functional and performance requirements and design criteria for any combination of parts, subassemblies, assemblies, units, groups, and sets[8] that performs a specific function and is essential to the completeness of a system or subsystem. The government subcategories of CI specifications are on page 59.

[8] Items joined together to perform an operational function.

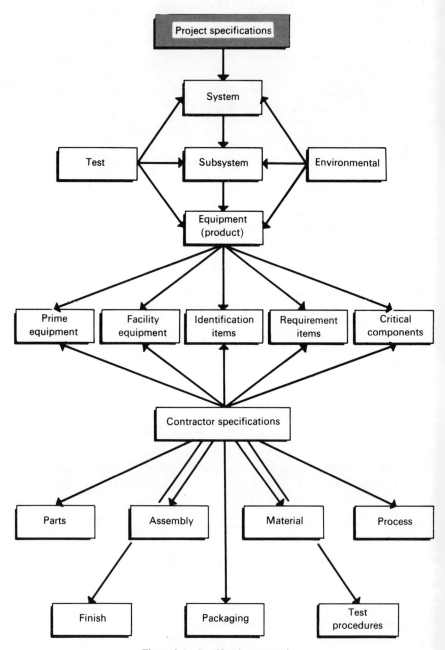

Figure 3.6   Specification categories.

MIL-STD-490 Terminology	AFSCM 375-1/NPC 500-1 Terminology	Description
Prime item	Prime equipment	Major operating complex units
Critical item	Engineering/Logistic critical component	Component within prime item that is technically or procurement sensitive
Noncomplex item	Identification item	Simplicity of function and design
Facility or ship	Facility	Structural, architectural features
Inventory item	Requirements item	Government furnished equipment

Format requirements for these categories are quite precise in MIL-STD-490 and therefore are not developed in discussion here.

4. *Interface Specification*

The interface specification establishes requirements for mating two or more parts, equipments, subsystems, or systems.

5. *Environmental Specification*

The environmental specification establishes the environmental requirements for qualification and acceptance testing. These specifications may be prepared at the assembly or system level, depending on system complexity and project requirements.

Contractor Specifications

The following specifications usually have broad company application rather than being in support of one particular CI.

1. *Parts Specification*

The parts specification establishes the design and detailed requirements for a piece or two or more pieces joined together as a part which cannot be separated without destroying the function of the part.

2. *Assembly Specification*

The assembly specification establishes detailed assembly procedures for any combination of parts, subassemblies, units, groups, or sets that performs a specific function.

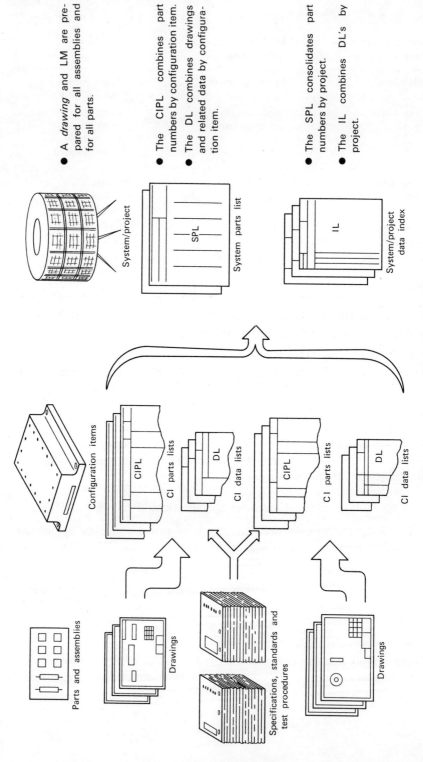

- A *drawing* and LM are prepared for all assemblies and for all parts.

- The CIPL combines part numbers by configuration item.
- The DL combines drawings and related data by configuration item.

- The SPL consolidates part numbers by project.
- The IL combines DL's by project.

Figure 3.7 Parts and data list interrelationships.

3. *Materials Specification*

The materials specification establishes the properties and detailed requirements for a raw or semifabricated material.

4. *Process Specification*

The process specification establishes material properties and detailed process control requirements for materials or items which require specific process operations, such as welding, heat treating, etc.

5. *Finish Specification*

The finish specification establishes the method and requirements for protective treatments and finishes for materials, parts, etc.

6. *Packaging Specification*

The packaging specification establishes the method and detailed requirements for packaging, handling, preservation, storage, or shipment of a part, material, etc.

3.6 PARTS AND DATA LISTS

The configuration summary records for the product may be best expressed by formatted data list information prepared by mechanized or manual means. Typical system requirements for what we have identified as an integrated record system are discussed in Chapter 2.

Briefly, the data listing capability must accept all formally released engineering data in great detail and interrelate usage data, indenture relationships, effectivity incorporation points, data cross references, quantity applications, and so on, into a complex central record from which parts and data lists may be responsively and accurately printed. (See Figure 3.7.) Typical of such outputs are the following parts and data lists.

Configuration Item Parts List

The CI parts list may be either an indentured or alphanumeric listing of all parts making up the CI. (See Figure 3.8.) Accordingly, the parts list will illustrate the indentured relationship, application, quantities, and authorized engineering data change letters for all parts and assemblies within the CI. Note that components or parts in the data bank (computer memory) may be suppressed to show assembly items only.

System Parts List

The system parts list is a tabulation of applications, cumulative quantities, and effectivity of all parts, components, and assemblies within a system. Figure 3.9 shows a typical system parts list.

Configuration Item Parts List

ABC Corporation
Washington, D.C.

Code Ident. **09876**   CIPL _____   Rev. _____

Project _____
CI Name _____   CI Number _____

Date _____ Sheet _____ Of _____

Item or Sequence Number	Change Code	Assy Level	Part Number	Description	Acct Rev. Ltr[a]	Effectivity		Next Higher Assy Part Number	Next Assy Quantity	Quantity Per CI	Remarks
						From	Through				

Preprinted format. Mechanized data lists may use computer-printed header and title information without preprinted guides.

[a] Acceptable revision of drawing for item at stated effectivity.

Note. Sequence numbers refer to address of data item-in computer.

Figure 3.8 Configuration item parts list, typical format.

ABC	System Parts List				09876 Code Ident.	SPL _____			Rev. _____
Corporation Washington, D.C.									
Project _____			Model _____					Sheet _____ Of _____	

| | | | | | | | Acct | Effectivity | |
Change Code[a]	Part Number	Item Name	Mfr's Code Ident.	Configuration Item Part Number	Next Higher Assy Part Number	Next Assy Quantity	Quantity per CI	Date	Rev Ltr[b]	From	Through

Preprinted format. Mechanized data lists may use computer-printed header information without preprinted guides.

[a] Change code: A = added; D = deleted; C = changed.
[b] Acceptable revision for item at stated effectivity.

Figure 3.9 System parts list, typical format.

ABC Corporation Washington, D.C.	Data List	09876 Code Ident.	DL _____	Rev.

Project:	Specification: _____ Model/Type: _____	Configuration Item Part Number _____ Name _____	Sheet ___ Of ___

Item Number	Document Size	Mfr's Code Ident.	Document Number	Rev.	Document Nomenclature	Remarks

Preprinted format. Mechanized data lists may use computer-printed header and title information without preprinted guides.

Figure 3.10   Configuration item data list, typical format.

ABC Corporation Washington, D.C.	Index List	09876 Code Identification	IL —————————	Rev.

Project	System Model/Type	Date	Sheet___ Of___

Item Number	Document Size	Manufacturing Code Identification	Document Number	Rev.	Document Nomenclature	Remarks
			Preprinted format. Mechanized data lists may use computer-printed header and title information without preprinted guides.			

Figure 3.11   Project data index list, typical format.

Configuration Status Accounting Report
Configuration Item Identification and Related Data Summary

Nomenclature: Stop Watch
Configuration Item: BA0001A
Specification: CM112100-1A 70 Sept 16
Drawing No: 4M00000 Watch Stop

As Of: 71 Nov 20
Contract Number: N000017-70-4444-A01
Contract Basic Line Item S/N: 0126
Contract Date: 70 Jan 28   Change: 68024AA

FSN: 0304z 2780-143-2901
Contractor: KB Watch Mfg Co
Part Number: 081079-4M00000-101

ECP Identifier	Action Priority	ECP Title	ECP Submittal Date	ECP Approval Date	Modification Work Order Number	Modification Work Order Date	Lot or Serial Number	Technical Publication Identification	Current to Date	Revision Schedule Update	Revision Actual Update
081079 0001- 00P	R	Change main spring	710404	710627	7727	710628	0129	15W-42-R01	710110	710215	710215
081079 0002- 00F	U	Reinforce stem. Add luminous dial and blue second hand	710622	710811	7910	710813	0200	15W-42-R03	710110	710915	710915

Date Code: 710915
Year
Month
Day

Sample accounting report
per MIL-STD-482.
There are many variations
in content, format, and degree
of detail, to be determined
in each case by the project
and customer. See Chapter 19.

Figure 3.12   Sample customer configuration status accounting report.

Configuration Item Data List

The data list is a tabulation of engineering drawings, specifications, and other reference documents required to fabricate and assemble a configuration item. (See Figure 3.10.) The documents are grouped in a specific order as follows:

1. Drawings.
2. Specifications and standards.
3. Government drawings.
4. Government specifications and standards.
5. Vendor data.

Index Data List

The index data list is a tabulation of data lists comprising a system or section of a system. Figure 3.11 gives a typical index data list.

Customer Configuration Status Accounting Records

Special customer requirements may necessitate the generation of customer-formatted configuration status accounting records. Part III, *Accounting*, discusses customer reporting requirements. A sample record is given in Figure 3.12.

3.7 TEST PROCEDURES

Test procedures establish detailed test objectives, test preparations, test equipment configuration, test precautions, nontest handling operations, and step-by-step test operations for obtaining data to be recorded within stated tolerances and accuracies using a specified configuration of test equipment. The general categories of acceptance and in-process test procedures are discussed in the following material.

Acceptance Test Procedures

The final acceptance test procedure defines the methods for verifying that the configuration item flight or operational equipment meets customer requirements. It defines the test conditions, gives step-by-step procedures for operational and environmental tests, provides pictorial test setups, and includes data sheets for recording test results. Exact test values and tolerances are given for evaluating equipment performance. The customer and quality assurance personnel witness the tests to assure that the correct procedures are followed and that test data were recorded accurately. In larger projects, the results of the tests are formally accepted or rejected by the project test review board (TRB). The configuration manager's representation on the TRB is usually mandatory.

Since final acceptance testing is the last step before delivering the product to the customer, changes to the procedure must be formally approved. Normally, CCB approval is not required for company acceptance of the procedure change and it is reviewed only by the project manager, design engineer, test engineer, and quality assurance engineer. The configuration manager coordinates the change package until it is approved and returned by the customer. If a customer's technical representative is assigned to monitor the project at the company, he may be authorized to approve procedure changes to minimize paperwork and delays.

In-process Test Procedures

In-process test procedures are engineering documents that are used to check out modules, printed circuit boards, subassemblies, and assemblies before final acceptance testing. They contain the same types of data that the final acceptance test procedure does, except in less detail and smaller scope. As a minimum, these procedures include applicable or reference documents, test requirements, test setups, and step-by-step test instructions. However, environmental tests are usually excluded except for checking equipment operation at high and low temperatures. In-process test procedures do not normally require formal control or customer approval because their main purpose is to assure the company that all components are working properly before final assembly and preparation for final acceptance testing, which is the basis for customer buy-off of the product. However, an informal system of controlling changes to these procedures is recommended. The control system can vary but should include the project manager, design engineer, test engineer, and configuration manager in the control loop. Special attention should be given to identification of in-process test procedures and the issuance of identification numbers. If data control is not responsible for issuing the numbers for these documents, then the configuration manager should issue them and maintain a log book of the title and number of each procedure. Accordingly, the configuration management office must be responsible for the identification, control, and status of both acceptance test procedures and in-process test procedures.

3.8   TECHNICAL OPERATING MANUALS

Product manuals describe operating instructions, installation, maintenance, calibration, and repair procedures and must be controlled to ensure their design compatibility at all times. Test set manuals must also be controlled to avoid the possibility of making changes that could result in erroneous performance measurements or damage to the product. (Test sets are used to check out the product for correct functional operation.) Because manuals are very important to the performance of the CI and the project, control of changes to these documents must be on a formal basis, although a somewhat different one from the drawing change control system. A system for changing manuals involves preparation of revised manual pages and a revision cover sheet, which includes approval signatures. The manual is a very important consideration in processing engineering change proposals to the customer. Also, it is essential that verification be made of the revised manual.

3.9   SHIPPING AND HANDLING INSTRUCTIONS

Depending on the care necessary in handling or installing the product, shipping and handling instructions or procedures may also require change control. These documents describe methods for unpacking, handling, carrying, and storing the product. Special precautions or potential hazards to personnel must be clearly and prominently presented in the text. In addition, sensitive or easily damaged parts or surfaces of the product must be identified and instructions must be given for avoiding damage to these elements. If gloves or special tools are required, they are described. The exact methods and degree of change control for shipping and handling instructions are determined by the project engineer, project manager, configuration manager, and quality assurance engineer.

Chapter 4

CHANGE CONTROL PROCEDURE

The following discussion covers a typical change control system and procedure for an equipment project. The detailed steps in the procedure vary depending on customer requirements and the complexity of the project.

The organizational description of the CCB and configuration manager responsibilities for change processing were discussed in Chapter 2. Typical change control forms for change processing are presented in Chapter 5. Chapter 19 describes the contractor and customer records for change status accounting and reporting.

A sample change control flow illustration is given in Figure 4.1 for Class I and II changes. Initiation of an engineering change is made by any one on the project by filling out a change request form (see Chapter 5) and submitting it to the configuration manager for review and preliminary approval. Upon the configuration manager's approval, the configuration administration and planning staff analyzes the change with respect to the following factors:

1. How many parts are due from vendors?
2. How many parts have been received from vendors? Where are they located?
3. How many parts are in stock and by which engineering orders have they been built?
4. How many parts are being fabricated in-house? In what stages of completion are they?
5. Does the part require test qualification? What is the qualification status?
6. On what models and projects is the part used? What are the next higher assemblies? If the part is an assembly, what parts that go into this assembly will be affected? Are clearances affected?
7. How many have been shipped? How many are spares?

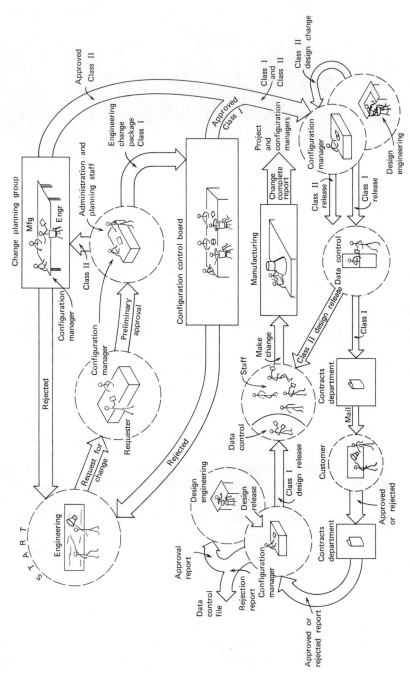

Figure 4.1 Configuration control cycle.

Based on the above information, the three-man change planning group (CPG) (configuration management representative, manufacturing representative, and design engineering representative) establish the effectivity of change with respect to the serial number of the first configuration item to receive the change and the disposition instructions for parts, assemblies, and materials affected by the change. Regarding changes where the immediate implementation of the change is not mandatory, such as improvement changes, the CPG determines the extent and cost of existing tooling, status of procurement, and the progress of open shop orders. The change point is set to minimize incorporation constraints. On the other hand, mandatory changes may require retrofit considerations with respect to delivered CI's (see Chapter 7) or major rework to in-process assemblies. Therefore the determination of the schedule and cost impact on manufacturing, procurement, and testing in responding to the mandatory situation must be carefully evaluated and recorded.

For this reason, major change packages (not limited to Class I) which represent the analysis and findings of the CPG are presented to the CCB. The change package is sent to the configuration control board members for review before the formal meeting is held. The configuration manager also prepares the agenda for the meeting and processes and coordinates all activities until the change is officially approved by the CCB and the customer, whereupon it is issued to data control. Concurrent functions of the configuration manager are to identify the change package[1] with a unique number and to prepare a status record for the change. A status record enables him to know where the change package is in the review and approval cycle. If the configuration control board approves the change, the configuration manager completes a configuration control board directive,[2] which records the results of the meeting, and forwards the change package to the customer through the customer's contract administrator. (The change package includes an engineering change proposal describing the change and its effects on cost and schedule.) When customer approval is received, the engineering design proposed by the ECP is completed by design engineering and released through the configuration manager to data control. The configuration administration and planning staff verifies that design documentation agrees with the planning determined during the review of the change.

Many routine as well as expedited changes will be submitted to the configuration manager as completed engineering orders, specification revisions, test procedure changes, and so forth, without prior project office evaluation. The role of the configuration administration and planning staff and the

---

[1] Not required in smaller projects.

[2] Engineering order is sufficient.

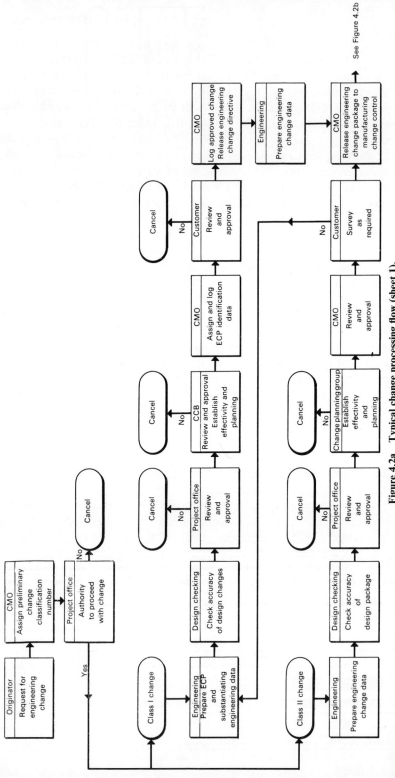

Figure 4.2a Typical change processing flow (sheet 1).

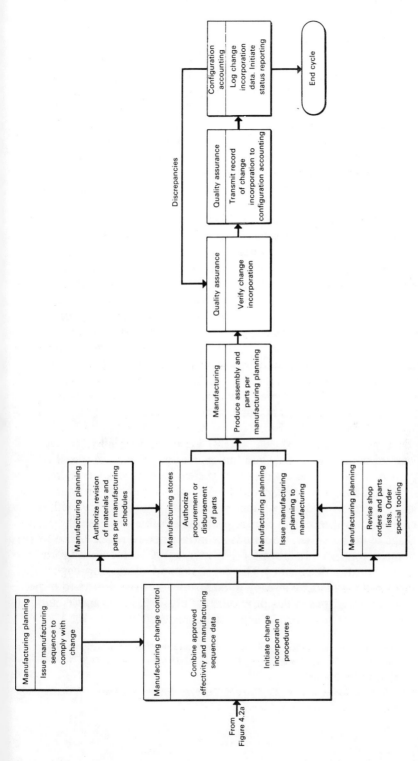

Figure 4.2b Typical change processing flow (sheet 2).

change planning group with respect to these changes will remain the same as to formally requested changes. The configuration management plan shall establish criteria for mandatory submittal of change requests versus completed engineering change documentation. These criteria are a function of time in the project cycle and are also dependent on project size and complexity. The use of formal engineering requests is not as necessary early in a project as it is later where test, production, and procurement commitments are well established.

No Class I changes are allowed to be made before written customer approval is received. A customer representative may be present on the CCB. His function is to provide preliminary customer review and approval of Class I changes and to review Class II changes to prevent erroneous classifications by the project manager or CCB.

In many cases a Class I change must be made immediately to avoid schedule slippage or excessive costs. Since the customer approval cycle may take two to six weeks,[3] the customer may be telephoned and the change described by the project manager or the customer's representative. If the customer agrees, official approval may be sent within a few hours to the company by teletypewriter exchange. If the customer's technical representative is authorized to conditionally approve a Class I change, then the change documentation authorizing its implementation can be released immediately. However, the change package must be sent to the customer for final approval and as a record of the change.

When a Class I change based on the previously listed areas—improved product performance, reduced weight, improved reliability, or reduced power consumption—is required, an engineering change proposal (ECP) is prepared. This document describes the engineering change, its purpose and value, additional costs, schedule slippage, and effects on other areas of the project, such as test equipment, manuals, and interfaces with other equipment or systems. The change proposal is submitted with the change package to the customer, who sends a contract change notice signed by the customer's contract administrator. (A letter, TWX, or preprinted form can be used for this document.) The exact method of making engineering changes will vary from company to company and from project to project. Thus many variations in this procedure are possible. For example, another illustration of a change processing cycle is presented in Figure 4.2.[4]

---

[3] DOD specifies that the company will receive contractual authorization within 24 hours for an emergency ECP; 15 days for an urgent ECP; and 45 days for a routine ECP. See Glossary for definitions of ECP types.

[4] Edited from J. Laine and E. C. Spevak, Teledyne System Company, "Configuration Management," *Space/Aeronautics*, November 1966.

Chapter 5

CONTROL DOCUMENTS

This chapter discusses nine basic control documents which may be most representative of those used in configuration control. These documents are only suggested as typical examples and are therefore not used for all projects. The complexity and detail also vary with the size of the project. Therefore these documents are presented as "middle-of-the-road" suggestions. Figure 5.1 shows the sequence of preparation of control documents.

1. Change request (CR).
2. Engineering change document (any of the following).
a. Advance design change notice (ADCN).
b. Design change notice (DCN).
c. Engineering order (EO).
d. Drawing revision engineering order (DREO).
3. Engineering change proposal (ECP).
4. Configuration control board directive (CCBD).
5. Specification change notice (SCN).
6. Deviations and waivers.
7. Contract change notice (CCN).
8. Engineering change follow-up report.
9. Stop order.

5.1 CHANGE REQUEST

The change request form, used by any organization to request product changes, should have company-wide availability. The originator of a change request provides the information required by the form (see Figure 5.2) and appends whatever supporting documentation is required, for instance marked-up drawings, sketches, and failure reports to facilitate and expedite review of the requested change. Completed change requests are forwarded

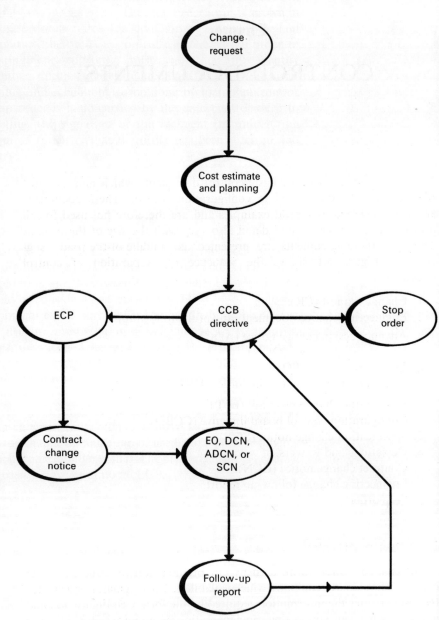

Figure 5.1 Sequence of control documentation preparation.

Change Request	Date	CR _____
	Revision	

Contract Number _____ CI P/N _____ CI Name _____

Item Affected (P/N) _____ Name _____

Next Higher Assembly P/N _____

Description of Change

Effectivity: S/N ___ Through S/N _____ Priority: ☐ Emergency ☐ Urgent ☐ Routine

☐ Production   ☐ Rework   ☐ Retrofit   Model Affected: Engineering ☐

Remarks: Prototype ☐

                                                                    Qualification ☐

                                                                    Operational ☐

                                                                    Spares ☐

                                              Class I ☐   Class II ☐

                                              Accepted ☐   Rejected ☐

Originator:	Configuration Manager:	Engineering:
_____ Date	_____ Date	_____ Date

Figure 5.2   Sample change request form.

Change Cost and Scheduling Estimate	Date	Number _____
	Revision	Project _____

Contract Number _____ CI P/N _____ CI Name _____

Change Request Number _____ Assembly Affected P/N _____

Models Affected _____ Name _____

Cost Estimate

Direct Labor	Burden Rate	Hours	Rate/Hour	Total Cost	Job Number

Materials

Item: _____   _____   _____

       _____   _____   _____

       _____   _____   _____

       _____   _____   _____

       _____   _____   _____

       _____   _____   _____

       _____   _____   _____

Travel: _____   _____   _____

       _____   _____   _____

Total Cost (includes G and A)   _____

Schedule   Go-Ahead Date _____ Completion Date _____

(Workdays from Go-Ahead)

                Drawings and EO's _____ Specification Revs. _____

                Test Procedures _____ Parts Received _____

                Tooling _____ Manufacturing Planning __

Planning and Control Manager	Date	Project Manager	Date
Configuration Manager	Date		

Figure 5.3 Sample change cost and schedule estimate.

by the originator to the configuration manager for preliminary approval action.

The configuration administration and planning staff reviews the change request, adds a change request number, records the number and title, and proceeds with the change evaluation and planning as illustrated in Chapter 4. If the requested change affects interfaces or is of such complexity as to warrant additional engineering analysis and detailed manufacturing and procurement inputs, an additional form or document may be necessary to record overall cost and planning impact. See Figure 5.3. This second form is completed by the change planning group for the approval of the configuration manager or configuration control board, as appropriate.

5.2   ENGINEERING CHANGE DOCUMENT

There are four common forms which may be chosen for changing engineering drawings: (a) advance design change notice (ADCN); (b) design change notice (DCN); (c) engineering order (EO); and (d) drawing revision engineering order (DREO). The objective of each is as follows:

1. *Advance design change notice* performs much the same function as the engineering order; therefore they may be considered one and the same. Their primary objective is to identify incremental changes to a particular drawing revision in "stepping" fashion and exist as changes "stapled" to the drawing until the drawing is revised, thereby incorporating the outstanding changes. The use of ADCN's and EO's is also faster and cheaper than revising the original drawing for each change.

2. *Design change notice* is written to incorporate outstanding ADCN's. In this respect the drawing revision engineering order serves the same function.

3. *Engineering order* has the same objectives as ADCN.

4. *Drawing revision engineering order* has the same objectives as DCN but applies to EO's.

5.3   ADVANCE DESIGN CHANGE NOTICE

Advance design change notice data (see Figure 5.4) include identification of the item to be changed, the revised drawing change letter (refer to Chapter 12), a unique identification number for the ADCN, a complete description of the change and its location on the drawing, the effectivity (see Chapter 7) of the change, other documents affected by the change, and instructions to staple the ADCN to the drawing or to incorporate the change into the drawing by redoing the original vellum (master).

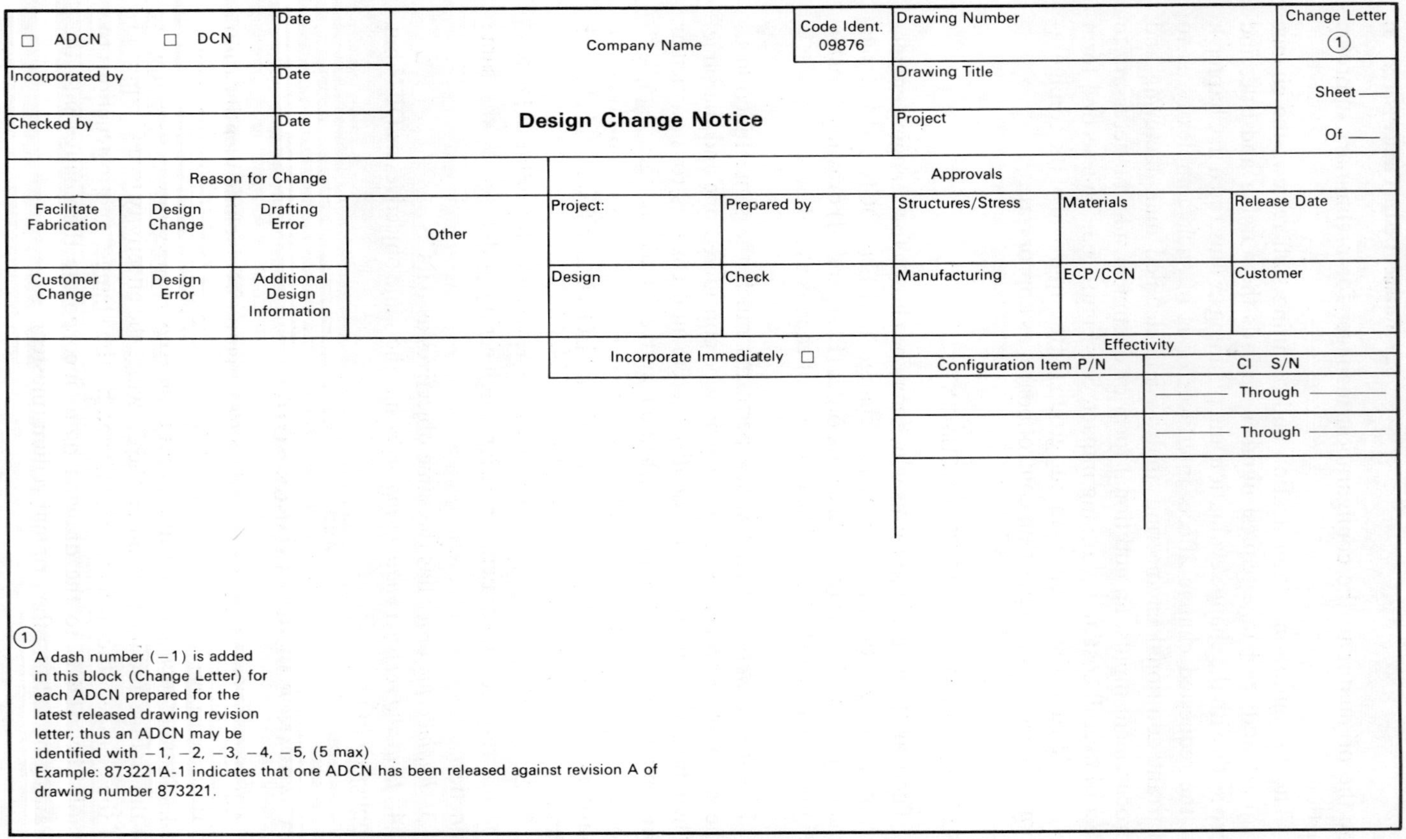

Figure 5.4 Sample (advance) design change notice: DCN, ADCN.

Form contents (Design Change Notice):

□ ADCN	□ DCN	Date		Company Name	Code Ident. 09876	Drawing Number	Change Letter ①

Incorporated by — Date

Checked by — Date

Design Change Notice

Drawing Title

Project

Sheet —— Of ——

Reason for Change:

| Facilitate Fabrication | Design Change | Drafting Error |
| Customer Change | Design Error | Additional Design Information |

Other

Approvals:

| Project: | Prepared by | Structures/Stress | Materials | Release Date |
| Design | Check | Manufacturing | ECP/CCN | Customer |

□ Incorporate Immediately

Effectivity

Configuration Item P/N

CI S/N

—— Through ——

—— Through ——

① A dash number (−1) is added in this block (Change Letter) for each ADCN prepared for the latest released drawing revision letter; thus an ADCN may be identified with −1, −2, −3, −4, −5, (5 max) Example: 873221A-1 indicates that one ADCN has been released against revision A of drawing number 873221.

As many as five[1] ADCN's may be written against a drawing before they must be incorporated into the original. Some agencies require outstanding (not incorporated) ADCN's to be incorporated every 30 days, even though only one or two ADCN's have been released and issued. When ADCN's are incorporated into the original drawing, the changes are briefly described in the upper right-hand corner of the drawing under the heading "Revisions." Once released, ADCN's are not revised but are superseded by a new ADCN that provides the correct data. All ADCN's are kept on file regardless of their acceptance or rejection by the configuration control board or customer.

5.4 DESIGN CHANGE NOTICE

A design change notice is written when the ADCN's are incorporated. The DCN, shown in Figure 5.4, is an official notification that the ADCN's have been incorporated. The DCN and the drawing are stapled together and released for distribution by data control. As shown in Figure 5.4, the same form can be used for both an ADCN and a DCN with the appropriate box marked to identify its type.

5.5 ENGINEERING ORDER

Engineering orders are used in some companies instead of ADCN's and DCN's (see Figure 5.5). A minor area of difference is that the EO requires a separate identification number while the ADCN and DCN use the drawing number. In other aspects the ADCN and EO are similar; that is, other data requirements are the same as for the ADCN. The EO is stapled to the drawing as the ADCN is. Engineering orders, like ADCN's, are not revised or identified with a change letter when a change is necessary. Instead, a new EO is written to cover a change in description, identification, or effectivity. EO's are also used as a stop order mechanism for shop or assembly work. (See Section 5.13.)

5.6 ENGINEERING CHANGE PROPOSAL

For many years engineering changes to government contracted equipment were handled in an informal manner. Sometimes an engineer and a piece of chalk were the only items required to make a change. However, with the increasing complexity of hardware, this method of change implementation became unworkable. As a result, changes that were beyond the scope of the contract had to be formally approved by the government. *ANA Bulletin* No. 390 contained some of the first guidelines for change control. This

---

[1] An arbitrary limit established to prevent attaching an excessive amount of EO's to a drawing.

ABC Corporation
Washington, D.C.

| Code Ident. 09876 | **Engineering Order** | Drawing Number | Sheet | Of |
| | | | Rev/EO No. | |

Type of EO

- ☐ Change
- ☐ Revise Drawing
- ☐ Expedited
- ☐ Class I
- ☐ Stop Order
- ☐ Deviation
- ☐ Limited Effectivity
- ☐ Class II

Disposition of Parts/Mat

	Use	Rework	Scrap	See Note
In Process				
Completed				
Assembled				

Reason for Change

☐ Mandatory   ☐ Record   ☐ Improvement

Drawing Title

Project	Next Assembly	Effectivity	
		CI Part No.	CI Ser. Nos.
Alpha-z	873211	802000-1	001 and Subs[a]

[a] Subs = Subsequent

| Originator | Check | Stress | Materials and Processes | Configuration Manager |
| | | | | Engineer |

Figure 5.5 Sample engineering order.

bulletin was superseded by *ANA Bulletin* No. 445 and more recently by MIL-STD-480, which defines Class I and Class II changes as being in or out of the scope of the contract. MIL-STD-480 requires that an ECP be written when engineering changes affect the physical, functional, performance, maintenance, or logistics characteristics or facilities, ground support equipment, and operational equipment, that is, any Class I change.

A sample ECP form (per *ANA Bulletin* No. 445[2]) is shown in Figures 5.6 and 5.7. Note that ECP's are used for changes to design requirements as well as for changes that occur during the development phase of the project.

The first data entries cover exact identification of the CI or part to be changed, the company identification code, priority ratings, production status of the equipment, and the contract specification.

A change is next described in sufficient detail for clear understanding. Detailed design analysis is not normally given but preliminary drawing schematics, illustrations, and test descriptions are included, or appended, when necessary for completeness. The reasons for proposing the change are itemized and may include superior performance, higher reliability, easier construction, lighter weight, and so on. Other data include the need for new tooling, tests, test fixtures; alternate solutions; production effectivity; and the estimated cost for making the change on production equipment.

Recommendations also may be made to rework (retrofit) an already completed product. If new requirements for servicing, maintaining, retrofitting, and documenting the product will result, these requirements are specified. On the second page additional cost and schedule requirements for the retrofit are given, and the other areas affected by the change are checked off in the appropriate boxes. The date for authorization to proceed is also given.

The production effectivity (block 10) will apply to the serial numbers of the CI, not to the part affected by the change. Production effectivity points proposed on a firm or estimated basis are established on the basis of the company's processing capability and the customer's approval reaction time as well as on engineering, procurement, and manufacturing leadtime considerations. For this reason, the company should qualify the recorded production effectivity point by indicating the date (see block 24) by which contractual coverage is required to meet the specified effectivity point. This date is shown in block 10 as follows:

SN 04 and on (based on receipt of a CCN
by 4/20/71)

---

[2] *ANA Bulletin* No. 445 is superseded by MIL-STD-480 (see Chapter 22) but the ECP form in 445 is given here because it is used more widely. However, the form in MIL-STD-480 should be used if a new form is being sought for adoption.

(ABC) Corporation Washington, D.C.	Engineering Change Proposal (prepared per *ANA Bulletin* 445)	Page 1 Of ___ Pages Date

	Manufacturing ECP Designation						
1	Model or Type Designation	Manufacturing Code	System Designation	ECP Number	Type	Rev.	Corrections

2	Contractor's Recommended Priority	☐ Emergency	☐ Urgent	☐ Routine	☐ Compatibility

	Contract Number	Contract End Item Nomenclature	Contract Specification	
3			Specification Number	Affected
			Drawing Number	☐ Yes ☐ No

	Name of Part or Lowest Assembly Affected	Part Number or Type Designation	In Production
4			☐ Yes ☐ No

	Title of Change
5	

	Description of Change
6	

	Justification for Change	
7		☐ Requested by Procuring Activity per: Ref:_____ ☐ Initiated by Contractor for: 1. Compliance with New or Revised Specifications___ 2. Fix for Unsatisfactory Report Numbers_____ 3. Other_____

	Developmental Requirements
8	

	Alternative Solutions
9	

	Production Effectivity
10	

	Estimated Cost for Change in Production
11	

	Recommendations for Retrofit
12	

	Retroactive Effectivity	SC/MWO/TCTO Required
13		⏜ See list of acronyms ☐ Yes ☐ No

Figure 5.6 Sample ECP, sheet 1.

Engineering Change Proposal (cont.)	Page __2__ Of _____ Pages
ECP Number	Date

14	Estimated Cost for Retrofit Kits Estimated Out of Service Time	Estimated Man Hours per Unit to Accomplish Operating Activity _____ Field Activity _____ Depot _____ Other _____

15	Estimated Kit Delivery Schedule	Estimated Cost of Special Tools	Estimated Delivery Schedule of Special Tools

16	Effect on Operational Employment ☐ Safety Incl: _____ Para: _____ ☐ Combat Effectiveness Incl: _____ Para: _____ ☐ Reliability Incl: _____ Para: _____ ☐ Service Life Incl: _____ Para: _____ ☐ Operating Procedure Incl: _____ Para: _____ ☐ Operating Installations Incl: _____ Para: _____ ☐ Radio Frequency Interference Incl: _____ Para: _____ ☐ Activation Schedule Incl: _____ Para: _____

17	Effect on Contract and Specification Requirements ☐ Performance Incl: _____ Para: _____ ☐ Weight, Balance, and Stability Incl: _____ Para: _____ ☐ Delivery Schedule Incl: _____ Para: _____

18	Effect on Logistic Support ☐ Individual Training Incl: _____ Para: _____ ☐ Training Installations Incl: _____ Para: _____ ☐ Maintenance Procedures Incl: _____ Para: _____ ☐ Overhaul/Rework Methods Incl: _____ Para: _____ ☐ Maintenance Installations Incl: _____ Para: _____ ☐ Maintenance Manpower Incl: _____ Para: _____ ☐ Nomenclature Incl: _____ Para: _____ ☐ Spare Parts Exhibit Incl: _____ Para: _____

19	Effect on Logistic Support Materials ☐ Spares Incl: _____ Para: _____ ☐ GFE/GFP ⎤ Incl: _____ Para: _____ ☐ CFE } See List of Incl: _____ Para: _____ ☐ AGE/SE ⎦ Acronyms Incl: _____ Para: _____ ☐ Trainers Incl: _____ Para: ☐ Data Publications Incl: _____ Para: _____

20	Other Considerations ☐ Interface Incl: _____ Para: _____ ☐ Physical Constraint Incl: _____ Para: _____ ☐ Operational Computer Programs Incl: _____ Para: _____

21	Target Completion Date and Summary of Estimated Total Program Costs

22	Development Status

23	Summary of Effect of Proposed and Previously Approved Changes on Major End Item

24	Date by Which Contractual Authority is Needed

25	

Figure 5.7 Sample ECP, sheet 2.

Configuration Control Board Directive	Date	CI Number _____
	Revision	Project: _____

Contract Number _____ CI P/N _____ CI Name _____

Change Request Number _____ Assembly Affected: P/N _____

Models Affected _____ Name _____

Approval Statement:

Serial Numbers Affected:

Planned Action Required	Cost Authority[a]	Organization	Schedule	Endorsement
Configuration Manager		Project Manager		

[a] Cost authority identifies job number to receive allocated costs for accomplishing the task.

Figure 5.8 Sample CCB directive.

The procedure for the detailed step-by-step completion of ECP forms is given in *ANA Bulletin* No. 445 and MIL-STD-480 and is therefore not discussed further here.

Note that ECP's may be revised when necessary and re-released. Unlike EO's, the original identification number may be retained and changed with revision letters. The ECP may also include a detailed cost backup sheet that identifies the types of labor categories required to make the change and the hours for each category. Categories include engineers, technicians, assemblers, inspectors, and wire men. The costs for materials and other items, such as special tools and test fixtures, are also included.

5.7 CONFIGURATION CONTROL BOARD DIRECTIVE

The configuration control board directive is a record of the control board meeting results and is signed and released by the CCB chairman. A sample form is shown in Figure 5.8. It is a summary of the information presented at the meeting and identifies retrofit, interface requirements, and items affected, such as spares, manuals, and specifications. The directive is also an official record of the CCB approval or disapproval of a proposed change.

5.8 SPECIFICATION CHANGE NOTICE

A specification change notice describes an approved change to an officially released detail specification for a product. This notice is shown in Figure 5.9. Approval signatures include all key engineering and project activities. The change is described in detail and the sheet attached opposite the old specification page. An SCN is prepared and delivered with each ECP that also affects the system detail CI specifications.

5.9 DEVIATION REQUEST AND AUTHORIZATION

Occasionally, during manufacture of an item, the project may need to deviate from certain specifications, drawings, manufacturing requirements, quality provisions, or test requirements because of special problem areas, the absence of specified materials or parts, or a change from normal procedures. When this is necessary, customer approval of deviation from the requirement is requested with the form in Figure 5.10. A customer-approved deviation allows delivery of the changed product without being contractually nonresponsive and without necessitating correction or revision of applicable documentation. A deviation differs from an engineering change in that an approved engineering change requires corresponding revision of the documentation defining the affected item, whereas a deviation does not involve

(ABC) Corporation		SCN _____
Washington, D.C.	**Specification Change Notice**	Page _____ Of _____
Code Identification 09876		Date _____ Superseding _____
EO/ECP Authority	Type, Model, Series	Specification Number
Contract	Contractual Authority	File Opposite Page Number
Effectivity	Approvals: Engineering _____ Configuration Mgr. _____ Quality Assurance_____ Test Engr. _____	

Figure 5.9   Sample specification change notice.

(ABC) Corporation Washington, D.C.	Code Identification 09876	**Deviation Request and Authorization**	Deviation Number
			Sheet Of
Project		Applied to Document	Date

Identification of Item or Operation Affected

☐ To be incorporated

or

☐ .Limited as Follows

☐ Eng Model S/N____
☐ Qual S/N _____
☐ Proto S/N _____
☐ Prod S/N _____

Nomenclature _____

Part Number _____ Rev Ltr ____ Serial Number _____

Manufacturing Operation _____ Mfr. order No. _____

Test Operation Affected _____

CI Part Number _____ CI Name _____

Deviation

Reason _____
_____
_____
_____
_____

Description _____
_____
_____
_____
_____
_____
_____
_____

Approvals

Engineering _____ Date ____

Quality Assurance _____ Date ____

Test. Engineer _____ Date ____

Configuration Manager _____ Date ____

Project Manager _____ Date ____

Customer _____ Date ____

Figure 5.10   Sample deviation request and authorization.

<table>
<tr><td colspan="2" rowspan="3">Design Change Completed Report</td><td colspan="2">DCCR (CCB directive number)</td></tr>
<tr><td colspan="2">☐ Qual ☐ Proto ☐ Ops[a] ☐ Spares</td></tr>
<tr><td>Date</td><td>Sheet Of</td></tr>
<tr><td>Item Affected P/N</td><td>Item Name</td><td colspan="2">Ref. ECP (where applicable)
Ref. CCN (where applicable)</td></tr>
<tr><td colspan="2">Change Summary Statement:</td><td>CI's Affected</td><td>Serial Number</td></tr>
<tr><td colspan="2">Change Incorporation Summary:
1.</td><td colspan="2">Incorporating Documents (engineering, manufacturing, test)
1.</td></tr>
<tr><td colspan="2">2.</td><td colspan="2">2.</td></tr>
<tr><td colspan="2">3.</td><td colspan="2">3.</td></tr>
<tr><td colspan="2">4.</td><td colspan="2">4.</td></tr>
<tr><td colspan="2">5.</td><td colspan="2">5.</td></tr>
<tr><td>Manufacturing</td><td>Quality Assurance</td><td colspan="2">Test Engineer</td></tr>
</table>

[a] Ops = Operational.

Figure 5.11 Sample design change completed report.

revision of the applicable specification or drawing. An approved deviation accompanies the CI as a part of the data package described in Chapter 7.

5.10 WAIVERS

Waivers are documents, similar to deviations, that allow the company to avoid complying with contract or specification requirements. Thus a waiver permits acceptance or use of the product when it does not meet specified requirements. The waiver must be approved by the customer. The main difference between the waiver and the deviation request is that a deviation permits a change from contract requirements before the change is implemented whereas a waiver permits a change after it has occurred. The waiver must contain all information necessary to describe the change and the reasons for it. No changes are made to the documentation describing the product when a waiver is issued. An approved waiver accompanies the product as a part of the data package described in Chapter 7.

5.11 CONTRACT CHANGE NOTICE

A contract change notice is a document from the customer approving a recommended change and authorizing an increase in project cost and a delay in the delivery of the CI. The change document (ECP, EO, ADCN, or DCN) is referenced by number and date on the CCN to assure that the correct change is made. A teletypewriter exchange or letter from the customer's contract officer or administrator can be used for a CCN. If no change in funding is permitted, this will be stated in the CCN. The customer's contract administrator signs the CCN and thereby makes it a legal contract document.

5.12 ENGINEERING FOLLOW-UP REPORT

The engineering follow-up report officially notifies the configuration control board that the change has been completed as directed by the EO, DCN, or ADCN. Before its distribution it is approved by the responsible engineer and configuration manager. A sample form for a follow-up report is given in Figure 5.11.

5.13 STOP ORDER

A stop order is prepared to notify manufacturing or testing to stop work on an item for the CI because an engineering change is in process. Its purpose is to prevent unnecessary expenditure of labor and materials on an item that will have to be reworked or scrapped at a later date. The stop order may

Estimated Date of Stop Order Lift: —————		**Stop Order**	SO Number —————
CI Serial Numbers Affected			Date —————

Part Number	Item Name	Next Assembly Part Number	CI Part Number

Reason for Stop:

Engineering Notes:

Responsible Engineer	Project Engineer
Other	Other

Manufacturing Notes:

Manufacturing Representative	Configuration Manager
	Project Office

Figure 5.12 Sample stop order.

effect a savings greater than the initial fabrication labor because rework may take more time than the initial effort. The stop order form is shown in Figure 5.12. The stop order requires the configuration manager's and project engineer's signatures before data control can release the order to manufacturing or testing. The stop order must clearly and fully identify the part and serial numbers affected. The estimated date for lifting the order is desirable for planning and scheduling to be done by manufacturing or testing. When the stop order is hand-carried by the data control clerk to manufacturing or testing, the signature of the recipient (manufacturing representative) is required to verify that the order has been received and read.

5.14   COMBINATION FORMS

The control document that describes a change is of critical importance in the development of data that ultimately describe the actions to be taken in the diverse areas that will be affected by the change. It is also desirable that the number of forms used be kept to a minimum to reduce costs and paperwork processing. Therefore, the consolidation of different forms into a single document should be considered whenever possible.

For example, a change directive may be used to handle both the request for a change and the ultimate authorization of that change (CCB directive) after it has become fully described. Thus the document (form) could be used for both pre- and post-CCB actions. The change directive develops like a rolling snowball. At first it is very small because the originator was probably only able to discuss the change in design without evaluating all its ramifications. By the time the change directive has passed through the CCB, however, many of the other consequences have been recognized and described.

It is most helpful if the change directive can be used and if all departments as well as the customer become familiar with it. The properly filled out change directive can then be easily converted into an ECP.

Chapter 6

RELEASE AND RECORDING PROCEDURE AND DRAWING CONTROL

Release and drawing control procedures are key areas of configuration control, for they set the pace and quality of the configuration management system. The key elements of the release and control system are (a) a data control facility for storing completed and approved documents released by the engineering and design group; (b) data control personnel for maintaining records of all documents, changes in documents and other related data, for controlling the distribution of documents to the project staff, and for preventing the unauthorized release of original documents to engineering, manufacturing, and test personnel; and (c) printing equipment for reproduction of drawings, procedures, and other documents.

Descriptions of key release and issuance operations follow. Detailed procedures may vary, but the fundamental system must employ these basic elements or their equivalents to provide adequate control. (Detailed requirements for release records are described in Chapter 15.)

6.1 RELEASE AND ISSUE FORM

A release and issue form is required for beginning the control procedure of a completed and approved document. This form is completed by the design engineer responsible for the preparation of the document and submitted with the original copy of the document to the data control clerk.

This form can be a simple one or one requiring the entry of considerable data related to the document. A sample form is shown in Figure 6.1. Most of the terms are self-explanatory but a few require definitions:

1. *NHA* refers to the next higher assembly drawing number into which the part defined by the drawing goes.

Engineering Release and Issue

Title		Prefix	Document Number	Change Letter

Job Number	Configuration Item/Model Number/Type	EID	Next Higher Assembly/Document

Date of Release by Engineering:	Anticipated Release Date	Required Release Date	Originator	Department Number

Size	Level or Grade	Distribution	Number of Sheets		Security

Checkboxes (left column):
- ☐ Formal
- ☐ Record Only
- ☐ Original
- ☐ Change
- ☐ EO, ADCN, DCN
- ☐

Checkboxes (middle column):
- ☐ Drawing
- ☐ Specification
- ☐ Procedure
- ☐ Parts/Wire/Drawing List
- ☐ Schematic
- ☐

Remarks and Limitations

Effectivity		CCB Directive	Release Authorized by	Date	Department Number
From	Through				

Type of Release

☐ Production ☐ Reference ☐ Limited ☐ Advance

Issued by	Issue Date

Figure 6.1 Engineering release form.

2. *Dist*, distribution of the document, is identified by a code number that refers to one of several kinds of distributions specified by the project office and maintained by the data control group.

3. *EID*, refers to end item designator, a number assigned to the equipment or a major subassembly that remains the same throughout the project.

4. *Level* refers to the grade (or form) of the drawing, which can be 1, 2, or 3 depending on its application. Refer to Chapter 3 for a description of drawing levels.

5. *Type of Release* indicates one of the four major types of release applying to documents. These are production, limited, reference, and advance. A production release concerns deliverable products and has the most stringent requirements; for example, formal drawing practices, complete review cycle, and tight change control. The limited release applies to a product that is not deliverable or to a deliverable engineering model that is used to check the form, fit, and function of the final flight or production CI. Reference releases are used for information only and do not affect manufacturing, procurement, or testing. An advance release applies to an item that requires a long leadtime to obtain from a vendor or supplier. Although the item's drawing has not been fully completed and approved, it can be released for the initial steps of procurement.

Data control is responsible for verifying that all blocks have been completed. When the completed form is verified, the document can be reproduced and distributed to the project staff. For tight control, the documents may be hand-carried and placed into controlled files in manufacturing and quality assurance departments. Note that without the approval signature of the originator and the configuration manager, no action can be taken on the issuance of the document. (The configuration manager signs his name in the "Release Authorized By" block shown in Figure 6.1.)

6.2   DOCUMENT STAMPS

Rubber stamps are used by data control to identify all drawings. At a minimum, all drawings are stamped with the releasing source identification (data control) and date. However, controlled documents are stamped with special identifiers besides. Controlled documents receive special care because they are used to build, test, or inspect flight products. When a revised document is released that supersedes previous documents, the data control clerk is responsible for replacing the old document with the new one to be sure that the latest change is received by the staff. Special stamps for these controlled documents may read as follows:

PRODUCTION—FLIGHT   (marking for drawing
delivered to manufacturing)

INSPECTION—FLIGHT   (marking for drawing
delivered to quality assurance)

These stamps indicate that the documents are for the manufacturing and quality assurance groups and that the CI built and inspected will be for flight applications. Other stamps may be used to indicate other end uses for the hardware, such as engineering model, prototype, tooling, test equipment, and vendor. A drawing that will be sent out of the company for fabrication of the item is stamped "Vendor Use Only." The documents having these special stamps are defined as "controlled" documents. Note that replaced controlled documents are either stamped "Superseded" and placed in the history file or destroyed by the data control clerk.

Documents can be reproduced by the data control clerk on the request of any project member. However, no official data control stamps may be added to the drawing without permission of the configuration manager. This permission is usually given by signing the release and issue form (Figure 6.1). Issuance of fabrication, testing, or inspection documents without the correct stamps should never be permitted by company management when right configuration control is required. This restriction prevents the inadvertent use of superseded or obsolete drawings that are no longer approved for use.

6.3 OBSOLETE DRAWINGS, PARTS LISTS, AND OTHER DOCUMENTS

Obsolete or superseded drawings, parts lists, and so forth, are eliminated from the data control active files and release records, except for a reproducible copy that is placed in the data control project history file. Thus, should any questions arise about past designs, a reference drawing can be produced for review and analysis by the project staff.

All documents put into the history file should be stamped "Obsolete" or "Superseded" over the title block or near the title block, depending on the availability of clear space. If documents are placed in paper folders or envelopes before filing, a different color, such as red, should be used to indicate that it contains an obsolete or superseded document. The outside of the folder or envelope should also be stamped "Superseded." Note that the terms "obsolete" and "superseded" are not synonymous. "Obsolete" denotes a document that is no longer used and has not been replaced by a revised version. "Superseded" applies to a document that has been replaced with a revised version and is therefore no longer a current document.

6.4 HISTORY RECORDS

History records include superseded drawings, design change orders, engineering change proposals, parts lists, wire lists, specifications, and manuals. Generally, only a reproducible copy is maintained by data control and no other groups are officially responsible for keeping these documents. However, the project office may keep duplicate records until the project is over. Upon completion of the project, data control will identify, box, and place these documents in an area set aside for storage. In some companies these documents are microfilmed to save space.

6.5 CHANGE STATUS RECORDS

Data control maintains current change status records for all released and issued documents. These records—EO and ECP status lists, configuration difference lists, and test procedure revision lists—are described in Chapter 19.

6.6 DOCUMENT RELEASE STATUS FILE

A document status file consists of a card (for example, 5 by 8 inches) for each document officially released to data control by engineering. This card, shown in Figure 6.2, is made out when the document is released by engineering to data control with its completed release and issue form. Key data on the card include:

1. Document title.
2. Document number.
3. Contract end item or CI number.
4. Next higher assembly.
5. Drawing size (A, B, C, etc.).
6. Change document number.
7. Drawing level or form number.
8. Distribution code letter or number.
9. Originator's name and department.
10. Document change letter.
11. Project title.
12. Job or project number.

A primary data entry on the card is to record the numbers of revision documents, such as ADCN's, SCN's, and EO's. When the release of the

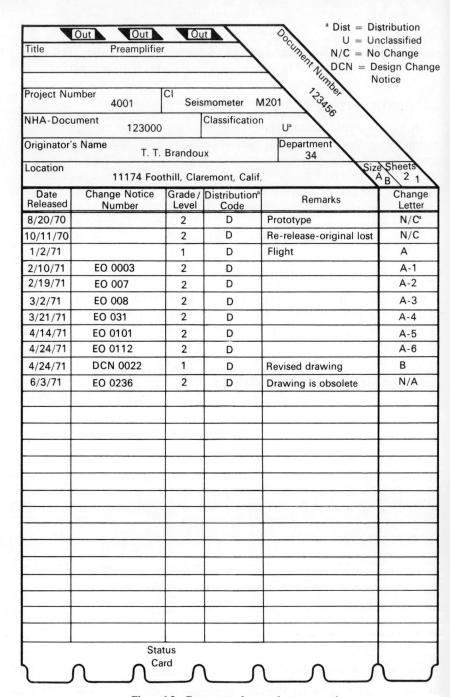

| | | | Out | | | | Out | | | | Out | | | | Document Number 123456 | [a] Dist = Distribution
U = Unclassified
N/C = No Change
DCN = Design Change Notice |

Title: Preamplifier

Project Number: 4001   CI   Seismometer   M201

NHA-Document: 123000   Classification: U^a

Originator's Name: T. T. Brandoux   Department: 34

Location: 11174 Foothill, Claremont, Calif,   Size A B   Sheets 2 1

Date Released	Change Notice Number	Grade / Level	Distribution[a] Code	Remarks	Change Letter
8/20/70		2	D	Prototype	N/C[a]
10/11/70		2	D	Re-release-original lost	N/C
1/2/71		1	D	Flight	A
2/10/71	EO 0003	2	D		A-1
2/19/71	EO 007	2	D		A-2
3/2/71	EO 008	2	D		A-3
3/21/71	EO 031	2	D		A-4
4/14/71	EO 0101	2	D		A-5
4/24/71	EO 0112	2	D		A-6
4/24/71	DCN 0022	1	D	Revised drawing	B
6/3/71	EO 0236	2	D	Drawing is obsolete	N/A

Status Card

Figure 6.2   Document release and status record.

original document is authorized, the person withdrawing the document must sign and date the card in the "Remarks" column. If a separate sign-out card is used, this card is signed by the person taking the vellum and the data control clerk files it with the document status card or file. When the original vellum is returned to data control, the name of the person who withdrew the vellum is crossed out with a single line to keep the name legible. All entries must be in ink to avoid inadvertently erasing historical information.

6.7   DRAWING AND RECORD CONTROL

Data control is responsible for the physical retention, protection, and control of drawings and records. Released drawing vellums are kept in files organized and identified for easy document retrieval. These files also protect vellums from damage. Vellums should never be folded and are filed flat whenever possible to prevent curling. When necessary, large drawings may be rolled and stored in special holders.

Engineers, designers, and draftsmen must not be allowed access to drawing vellums that they could remove, change, and return without proper identification, evaluation, and approval of the changes made. Therefore, a separate facility, closed off from the rest of the company, is necessary to ensure the integrity of the information on drawing vellums. When a staff member needs a vellum, he must present a properly completed form signed by the configuration manager that authorizes data control to release a specific vellum for inspection or revision. A design change notice or DREO can be used for this purpose.

Like drawings, prime records should not be available for use by the engineering staff. Instead, they should be kept in data control, usually in open tub files, for easy reference and updating by data control clerks. Release and issue forms and document status cards should be kept in separate files. Both forms and cards are filed, using the applicable document identifier, in numerical order by project.

Chapter 7

OTHER MAJOR ASPECTS OF CONFIGURATION CONTROL

Configuration control responsibilities cover an additional group of areas not described in the preceding chapters. These areas are essential to accurate and complete control of the product configuration and are therefore grouped and briefly discussed in this chapter.

7.1 CONFIGURATION ITEM SELECTION

The principal objective of the early definition phases of a project is to "allocate" the system or product into manageable items of hardware, data software, and/or operational computer software. These manageable tasks (configuration items) become, through discrete engineering descriptions and in conjunction with the project work breakdown structure, a product oriented framework for such project planning and management tasks as the following:

1. Planning the accomplishment of the proposed work.
2. Establishment of proposed hardware schedules.
3. Determination of engineering, manufacturing, procurement, and test budgets, and cost progress.
4. Delegation of product development responsibilities through the work breakdown structure.
5. Application of configuration management principles for identification, control, and accounting of product status.

In considering the selection of configuration items, the following criteria are suggested:

1. Item is preidentified by the buyer as a CI.

2. Item has a specification or a top assembly drawing, or both, subject to buyer approval.

3. Item will be maintained and operated as a separate entity and therefore may be allocated to different locations.

4. Item provides major support to a parent CI and interfaces directly with other CI's.

5. Item may be procured in the assembled condition as a spare.

6. Item is subject to separate qualification and/or acceptance testing by contract with resulting recording of test data.

7. Item may be procured on a major subcontract which includes the design responsibility.

8. Item is deliverable test equipment, special test equipment, in plant test equipment, special tools, or facilities, etc.

9. Item is government furnished equipment (GFE) to be incorporated in the system or product.

7.2 NOMENCLATURE

Nomenclature refers to a set of names or symbols given to items or parts of a product or CI as a means of classification and identification. The important thing to be observed is the consistency in the use of the titles given to various items of the product. The use of different titles for the same item can cause confusion among engineering, manufacturing, inspection, data control, drafting, quality assurance, and publications and writing personnel, as well as to the customer.

Product titles are usually provided by the customer. However, drawing titles for items that go into the CI usually are based upon titles given by engineering. Many projects use the "Federal Item Identification Guide for Supply Cataloging, Handbook H6-1" for deriving drawing titles. The key rules for deriving titles specified by this handbook are the following:

1. Find the best noun for the item in Part 1 of the handbook.

2. Verify that the definition given for the name basically describes the item on the drawing. If it doesn't, find another noun name that better describes the item.

3. Find a noun name modifier that best supplements the basic name; for example, bracket, *angle* or circuit, *gate*.

4. If items to be named are identical to a previously named item but have been modified by adding a bracket, add the modifier "bracket" to the name of the new item.

The configuration manager is responsible for implementing the standards for name selection and reviews drawing and document titles for correct application and consistency. However, the design and drafting group selects the names for the drawings as they are prepared. Conflicting noun names and modifiers are eliminated by the configuration manager whenever they occur.

7.3 SERIALIZATION

Serialization is the process of uniquely identifying each part, subassembly, and product in a family of products being supplied by the project. Serial number blocks are assigned by the configuration manager. Manufacturing draws on these assignments and is responsible for marking the hardware correctly and legibly with the applicable number. The quality assurance engineer is responsible for verifying that the serial numbers have been properly marked on the parts. Special care should be taken to place the serial number on the part so that it can be read after its installation into the next higher assembly, such as a printed circuit board. When necessary, the quality assurance engineer will add serialization .requirements on the purchase order and manufacturing instructions. The location and method of marking are given on the part drawing, or the drawing may contain a reference to a special procedure that should be used. When a part is too small for serialization, a tag with the serial number is attached to the part or accompanies the part in a plastic bag until it is completely installed in its next higher assembly. When the part is installed, the part and serial numbers are recorded into the manufacturing data related to the next higher assembly containing the part. Further discussion of product and part serial numbers is given in Sections 9.4 and 11.2.

7.4 LOT CONTROLLED ITEMS

A lot refers to items built of the same material, at essentially the same time, and by the same process so that all the parts have the same characteristics. Materials or parts lots are usually traceable to a production period of a few weeks. Depending on contract or company requirements, the supplier may be required to keep traceability data on all materials and parts shipped to the company for several years. The quality assurance group on the project is primarily responsible for lot control, data recording, and filing. However, the configuration manager is also responsible for monitoring the quality assurance system and for keeping records for complete configuration control and identification. (A "lot" is also referred to as a "batch.")

7.5   TRACEABILITY[1]

Traceability is the characteristic of a product and its parts that enables the parts to be tracked down to their source or origin and date of manufacture. Thus when a product failure occurs and the part that failed is analyzed and found to be an inherently defective part, the project manager can trace the part to its source. All parts that were produced by the vendor at that time can be identified, traced to their current use or application, and, if necessary, replaced with more reliable parts.

The configuration manager is responsible for ensuring that procurement, inspection, quality assurance, manufacturing, and engineering are following the system established at the beginning of the project for recording the origin and present location of all parts, materials, modules, and subassemblies.

To achieve reliable traceability, each supplier of CI parts as well as the company must keep identification records for each item. These records include the dates of manufacture, date of receipt by the company, test data, parts failures or defects, and identification numbers. These records provide a complete history of the part or material from the date of manufacture to the final installation and testing of the subassembly which it goes into.

Each part and subassembly is identified with a serial number. A purchase order number may be used for materials that cannot be serialized. These material and part numbers are marked on each item and recorded on a parts list or kit list. (A kit list is a material control document that identifies each part that goes into a subassembly or assembly.) This parts list and a manufacturing order, which describes the parts installation and assembly procedure, accompany the part through final assembly. All data pertaining to lot numbered parts, such as test results, inspection records, and manufacturing records, are recorded by the quality assurance representative, and a copy of this identification information is issued to the configuration manager if he requires this data in his file. However, inspection department files may be sufficient for his needs.

A controlled storeroom is essential for maintaining traceability and avoiding parts damage or loss of identification. This storeroom contains thousands of small bins for storing parts after they are inspected (for damage, quantity, and QA requirements) and approved by the inspector. The parts are logged in by the storeroom clerk and carefully stored in bins that are marked with the part number and name. The clerk keeps accurate books on the status and quantity of each part and he must be sure that all parts are identified. Another major function of the storeroom clerk is to assemble the parts into kits based on approved kit lists, prepared by the

---

[1] Traceability applies to more than just parts; it applies to engineering, manufacturing, and test documentation as well.

materials coordinator, that identify all the parts and materials required to make each module, printed circuit board, or subassembly in the CI. Completed and inspected kits, kept in sealed boxes, are released to the manufacturing engineer when he is ready to begin assembly of the parts.

To assure accurate traceability, the controlled storeroom must be kept locked whenever the clerk is not there; unauthorized persons should not be allowed into the room at any time. In addition to the clerk, only the project manager, materials coordinator, and quality assurance engineer are usually authorized to enter the room.

7.6   EFFECTIVITY

Effectivity is that point in the manufacture of a CI at which an engineering design change is introduced. Change effectivity applies to all CI's scheduled for manufacture, including spares. It is the responsibility of the configuration manager to determine the effectivity of each change. Effectivity data may be recommended to the configuration manager by the designer, draftsman, unit or production (manufacturing) engineer, and so on; however, it remains the final responsibility of the configuration manager to insure that these data are correct.

Effectivity is expressed in terms of the part and serial numbers of the next higher CI's affected, unless the change is to a drawing of a CI, in which case the effectivity is expressed in terms of the part and serial numbers of the CI itself. That is, the change looks to successive next higher assemblies until the first serialized CI is reached. Changes involving two or more CI's must provide separate effectivity for each. Effectivity is not expressed in terms of the serial number of the component or assembly affected because of the difficulty in controlling change implementation at this level. Thus effectivity is always applied to the CI serial number level.

7.7   PART DISPOSITION INSTRUCTIONS

Part disposition instructions, which must be compatible with change effectivity, are based on technical judgement about the feasibility of the disposition, not on actual part status. That is, can a part be reworked to its new revision level or part number configuration, not are there parts in existence.

These instructions document what is acceptable from an engineering standpoint. No less restrictive disposition may be used; however, manufacturing may elect a more restrictive disposition if it would be advantageous. (For example, if "Use" is indicated, parts may be either used, reworked, or scrapped; but if "Rework" is indicated, they may be reworked or scrapped, but not used.) Since the indicated disposition reports what is acceptable to engineering, it establishes this portion of the acceptance criteria for inspection as well.

Unless a stop order has been written, disposition instructions shall be made as if parts have already been fabricated, even when it is known that no parts exist.

"Use" Disposition. A "Use" disposition is indicated only when (a) engineering considers the change to be interchangeable and (b) engineering does not care at what point the change is incorporated in the parts concerned. That is, parts either with or without the change are acceptable.

"Use" changes are normally made to improve manufacturing operations (for example, to specify a more efficient process, change a plating call-out for better producibility, and so forth); therefore they may be phased into the production cycle at the earliest convenient point. The effectivity of these changes shall be "Scheduled by MFG Operation."

"Rework" Disposition. A "Rework" disposition is used when parts reworked to the change described in the engineering order are acceptable to engineering. When "Rework" is specified, the use of existing parts without change is prohibited; however, existing parts may be scrapped, at the discretion of manufacturing, if this is more practical than reworking. (Reworking some assemblies may cost more than building new ones with new components.)

"Scrap" Disposition. A "Scrap" disposition is used when parts cannot be reworked to the change described in the EO and when reworked parts will be unacceptable from an engineering point of view. When "Scrap" is specified, the use or rework of existing parts is prohibited.

7.8 INTERCHANGEABILITY

When two or more items have the same functional and physical characteristics so as to be equivalent in performance and durability and capable of being exchanged for each other without changing the items themselves or adjoining items except for adjustment, and without selection for fit or performance, the items are interchangeable.

When design changes make an item not interchangeable, the item, its next higher assembly, and all progressively higher assemblies are reidentified up to and including the assembly where interchangeability is reestablished. Note that under no circumstances should noninterchangeable items have the same identifying number. Refer to Chapter 12 for further discussions on part number reidentification.

The project and configuration managers are responsible for ensuring that the above principles are used during the project. However, the project manager is primarily responsible because he usually knows when a part is not interchangeable before the configuration manager does.

7.9 SUPPLIER CONFIGURATION CONTROL

Many parts and components for a complicated CI are built and tested outside the company responsible for the overall project. Suppliers of these parts must use a configuration control system that is the same or similar to that applied to the project at the company's facilities. Therefore, the company must impose the applicable requirements on its suppliers and must monitor each supplier's configuration control system to ensure conformance to the requirements. The configuration manager, procurement officer, and quality assurance engineer work together to direct each supplier in implementing the required configuration control procedures.

Changes made in the supplier's product or design must be controlled by the company. The exact system does not have to be the same as that used by the company but all Class I engineering changes should be approved by the project engineer, reliability engineer, quality assurance engineer, and configuration manager. A formal CCB meeting can be held to review the change, or an informal review can be conducted by sending each CCB member a description of the change and why it is necessary. The supplier should be required to send Class II changes to the project manager for review and agreement to the classification. Usually the project manager's agreement to the change class is the only one necessary. If he has any doubts, however, additional opinions should be obtained from the appropriate project members.

The project and configuration managers and quality assurance engineer should also verify that the supplier is properly identifying each part with its own part and serial numbers. Parts too small for identifiers or serial numbers should be identified by lot number, traceable to records maintained by the supplier.

The company may impose drawing requirements on the supplier or may allow him to use his own standards, if they are acceptable. The supplier will also be required to have a controlled drawing change procedure, which ensures that revised drawings are properly identified and that the changes are clearly described in the space on the drawing for revision descriptions.

When suppliers do not have a configuration control system, training in control procedures is necessary. This training may be handled by the configuration manager and quality assurance engineer.

7.10 SPARES

Spare parts, modules, or subassemblies may be required as replacements for failed or worn out parts of a delivered CI. Part identification, change control, serialization, testing, and other related configuration requirements apply to these parts just as they do to the flight or operation CI. When these parts are used, the configuration description[2] of the CI must be changed to include the new serial numbered part.

Selection of the types and quantities of spares is an engineering function that depends on the lifetime and reliability of the parts used for the CI.

7.11 RETROFIT CONTROL

Control of each retrofit to a CI already delivered to the customer is essential to maintaining current descriptions of the configuration at the manufacturer's facility. Without control and status accounting, the exact configuration of the CI cannot be determined after a few retrofits have been made. And as time passes, even one retrofit can result in the loss of the exact configuration of the CI. It is therefore especially important that the configuration control system and procedures describe the exact method for initiating, documenting, incorporating, and reporting all retrofit work to an equipment already delivered to the customer.

When an equipment is retrofitted as a result of a design change, a complete description of the new configuration must be made defining the changes, such as new parts, wiring, finish, materials, tests, and so forth. The retrofit data package usually includes a special drawing, parts kit, assembly instructions, test requirements, and a retrofit completion report. The records for these changes must be incorporated into the record package, or log book, that is kept with the CI. In addition, the retrofit completion report, which identifies the equipment, or CI, by nomenclature and serial number (see Chapter 15), location, person making the retrofit, retrofit identification number, and date, must be sent to the company for the project history file.

7.12 DRAWING FORMAT

Drawing format will depend on the contract requirements. In many cases the format given in the company's design and drafting manual is satisfactory. However, government standards or specifications may be imposed. These documents, specified by the government, are usually the following:

---

[2] As-built list described in Chapter 19.

1. MIL-STD-100, "Engineering Drawing Practices."
2. MIL-D-1000, "Drawings, Engineering and Associated Lists."
3. MIL-D-70327, "Drawings, Engineering and Associated Lists."

The format requirements include drawing sizes, line weights, lettering size and style, title block, revision block, and next assembly block.

It is the responsibility of the configuration manager to review the format requirements and to verify that the drafting group is following the requirements of the customer.

7.13   ACCEPTANCE DATA PACKAGE

Upon acceptance of each CI, a data package is delivered to the customer along with the product. The configuration manager assists in its preparation, but the quality assurance engineer or documentation engineer may be responsible for compiling the package. This package contains all the data necessary to describe the final configuration of the delivered CI, including key drawings (such as the top assembly, subassemblies, schematics, wiring diagrams, and block diagrams), operating time records, parts replacement records, weight and center of gravity of the CI, verified configuration record, final detail specifications, acceptance test procedures and test data, a list of open (uncompleted) items, a shortage list identifying missing parts or assemblies, nonconformance reports, deviations, waivers, and failure reports.

To identify these data and to simplify their collection, a matrix is prepared as shown in Figure 7.1. This matrix identifies each data item required, depending on the CI classification—engineering model, prototype, or flight. The data can be compiled into log books or by other means depending on the convenience of the staff or on customer requirements. A checklist is also used and is prepared for the ADP as shown in Figure 7.2.

Unaccomplished Tests and Inspections

Unaccomplished tests and inspections include final acceptance tests and inspections not performed before delivery of the CI. The list identifies the tests or inspections not completed and includes planned dates for completing this work. The tests or inspections not completed should be identified by paragraph and page number of the source document. The reasons for lack of completion can be briefly described in the "Remarks" column. The form for recording unaccomplished tests and inspections is shown in Figure 7.3.

Shortage List

A shortage list is delivered with each ADP. This list identifies hardware or documentation not shipped with the product. The planned shipping

Contents of CI and Test Set Acceptance Data Packages

Deliverable Hardware \ Package Contents	Check List	Test Plan	Drawings — Top	Module/Assembly	Schematic	Wiring Diagram/Lists	Manual	Flight Specification	Equipment Log — Operating Time Record	Verified Configuration Record 6	Acceptance Test Procedure	Acceptance Test Data	Failure Reports	Unaccomplished Tests and Inspections	Shortage List	Open Items	Inspection Records 10	Weight and CG	Qualification Status List
Engineering models (S/N 01 and 02)	X	X	X	X [1]	X [2]	X [3][4]									X	X		X	
Prototype models (S/N 03 and 04)	X	X	X	X [1]	X [2]	X [3][4]			[7]	X	X	X	X	X	X	X	[8]	X	X
Flight models (S/N 05—S/N 12)	X	X	X	X [1]	X [2]	X [3][4]	X	X	[7]	X	X	X	X		X	X	X	X	X
Test sets (3)	X	X	X [5]			X [11]	X		[9]		X	X			X	X			
Astronaut trainer			[5]				X											X (weight only) [5]	

Notes

1. Includes assembly drawings for each of three modules (chassis assemblies, also referred to as bilvets)
2. Includes schematic for each of three modules
3. Includes wiring list for each of three modules
4. Includes overall wiring interconnect diagram
5. Included in manual
6. Drawing and revision list if it covers end-item
7. The elapsed time meter on CI will be used as a record of operating time. No other record is required
8. ABC corporation. Inspection records only
9. The elapsed time meter on the test will be used as record of operation time. No other record is required
10. Includes MO's, IR's, MRR's, MRB's, and waivers
11. Wire list

Figure 7.1 Sample acceptance data package contents matrix.

Check List
for
Equipment X Acceptance Data Package

Item Number	Title	Number/ Remarks	Already Submitted	Submitted With S/N	To Follow Delivery
1.	Checklist				
2.	Test Plan				
3.	Drawings				
a.	Top Assembly (TA)				
b.	Module 100 TA				
c.	Module 200 TA				
d.	Module 300 TA				
e.	Schematics				
	Module 100				
	Module 200				
	Module 300				
f.	Interconnect Diagram				
4.	Wire Lists				
	Module 100				
	Module 200				
	Module 300				
5.	Instruction Manual				
6.	Flight Specification				
7.	Verified Configuration Record				
8.	Operating Time Record				
9.	Acceptance Test Procedure	17069-ATP			
10.	Test Data	17069-ATP-DS			
11.	Shortage List				
12.	Open Items				
13.	Tests and Inspections				
14.	Inspection Records				
15.	Weight and CG				

Figure 7.2 Checklist for acceptance data package.

Date ______________
W.A. 17069

Unaccomplished Tests and Inspections
for
Equipment X-1, SN05 Acceptance Data Package

The tests or retests required for the X-1 before delivery to ABC Corporation but not completed are listed below:

Item Number	Test, Retest or Inspection Description	Assembly Affected	Planned Completion Date	Remarks

Figure 7.3 Unaccomplished tests and inspections for acceptance data package.

date for these shortages is also used to identify all assembly or manufacturing operation shortages that exist at the time of CI acceptance by the customer. A contract change notice approving shipment of the CI with these shortages should be obtained from the customer before delivery. A copy of this notice should be included with the shortage list (Figure 7.4).

Open Items

Open items, work or tests not completed on the CI, are identified on an open items list. The tests listed here are in-process or types of tests other than final acceptance tests. Unaccomplished final ATP tests are listed in the unaccomplished tests and inspections list mentioned above. The date of completion is given on the form shown in Figure 7.5. In addition, a brief description of the open items and any remarks pertaining to these items are included, as well as any approved engineering changes that have not yet been incorporated. As with the shortage list, a contract notice authorizing shipment of the CI with this uncompleted work should be obtained from the customer before shipment.

Qualification Status List

The qualification status list (QSL) identifies all the parts that are qualified for use in the CI and gives the specific basis for this rating. The reliability engineer is responsible for preparing the QSL. Since the QSL includes all the parts used in the CI, it may be over 100 pages long. The QSL contains the following types of data for each part listed as qualified:

1. Item nomenclature.
2. Environment.
3. Parameters.
4. Stress level (contract or specification requirement).
5. Verification of stress level capabilities (agency, location, document reference, and date).
6. Remarks.

Configuration and Serialization List

The configuration and serialization list may also be included in the data package. This list identifies deliverable items of the product by nomenclature, part number, and serial number down to and including the lowest inter-changeable repairable assembly level. This is a key record and great care should be taken to ensure its accuracy and completeness. The manufacturing engineer, quality assurance engineer, and configuration manager should verify the list's correctness and affix their approvals. [The configuration and serialization list is also referred to as the verified configuration record (VCR) or the as-built list.]

Date ______
W.A. 17069

Shortage List
for
Equipment X-1 SN05 Acceptance Data Package

(to be included with DD Form 250)[a]

The following items are not included with the delivered hardware and will be shipped separately on the dates shown below:

Item Number	Description	Ship Date	Remarks

[a] DD250 is a government form used to inspect and accept an equipment.

Figure 7.4 Shortage list for acceptance data package.

Open Items
for
Equipment X Acceptance Data Package

The following work was not completed on X-1, SN05 as planned and will be completed
on the dates shown below:

Item Number	Description	Due Date	Remarks

Figure 7.5   Open items list for acceptance data package.

Failure and Replacement Record

The failure and replacement record lists all failures and replacements, to the serialized assembly level, that occur during testing. Data recorded include the part and serial numbers, failure report number, time and cycles to failure, and a summary of the failures and disposition (rework or rejected) of the failed part. The quality assurance and reliability engineers are responsible for collecting the data for this record and for verifying its correctness. This responsibility includes monitoring testing and manufacturing personnel to ensure that they keep complete records of all rework to the CI.

Configuration Acceptance Certificate

This document certifies that the CI under acceptance test conforms to its interface control document, drawings, and specification. It should be signed by the project manager, quality assurance engineer, and customer representative. The certificate should be placed immediately after the data package title page.

PART II IDENTIFICATION[1]

[1] Parts of Chapters 8, 10, 12, and 15 have been adapted from NASA's NHB 8040.2, "Apollo Configuration Management Manual," AFSCM 375-1, "Configuration Management During Definition and Acquisition Phases," and MIL-STD-100, "Engineering Drawing Practices."

Chapter 8

IDENTIFICATION REQUIREMENTS

Accurate and complete identification of hardware and data required to design, build, test, and accept equipment is critically important to successful configuration management. Without adequate identification, we cannot hope to deliver equipment that meets the customer's requirements for performance, configuration, and data that are compatible, interrelated, and consistent. The identifiers assigned to equipment and data interrelate all significant technical and administrative actions that occur during a project. Administrative actions cover contract changes, authorizations for revising data, procurement of spares, schedule changes, and identification of persons initiating and processing technical changes. By interrelating technical and administrative actions, all project elements are tied together into a uniform, integrated package that allows the customer and company to trace hardware and data actions to their sources and authority for implementation. Interrelations among equipments and data are provided primarily by the CI number, CI nomenclature, contract number, part numbers, serial numbers, and product specification number.

The key requirements for successful configuration identification are the following:

1. All data and equipment items must be identified.
2. One individual or group is responsible for issuance and control of all configuration identifiers.
3. A system is established with written procedures for controlling identification operations.
4. An independent control group is established to verify that correct identification procedures are being followed.

An inadequate identification system can easily cause problems in identifying the CI composition. For example, if a piece part too small to be marked with its part and serial numbers is not otherwise marked, the customer and company will lack the identification data necessary to trace the part from the product containing it to the original vendor's lot or serial number. Thus, should the part from an unknown lot delivery be found to have a defect related to its original manufacture, it will be impossible to identify the other parts that were in the lot. Consequently all parts with the same item identification and vendor would have to be replaced in each product built during the project. If the lot number were known, only products containing parts from that lot number would have to be replaced. The cost and time to replace 12 parts as compared to perhaps 200 (all the equipments instead of three or four) would obviously be much less. To avoid this situation, each part that cannot be permanently identified on its body is placed in a transparent plastic bag or box that includes an identifier card containing the part number, nomenclature, serial or lot number, and the purchase order number. These data are retained until the part is installed on a higher level assembly, at which time the data are added to the kit list of the higher assembly. The kit list provides a permanent record for later use in traceability actions requiring the part lot or serial numbers.

The simplest aspect of identification requirements is that of physical identification. Items can be identified with nameplates, engraved lettering, gummed labels, tags, felt pen, or other methods. Some companies have resorted to the use of colored dots made with a paint brush to differentiate items, reliability levels, or lots of small components such as might be found in the electronics industry. Data or documents usually contain all their key identifiers on their title pages, as is usually required by the contract. The most difficult facets of identification requirements are the determination, assignment, and control of identifiers. Therefore, the following sections are devoted primarily to these areas.

8.1 GENERAL REQUIREMENTS

The discussions in the following sections provide the basis for satisfying the customer's requirements for identifying and interrelating the product and data to be formally accepted by the customer. Only six types of identifiers are necessary to meet these requirements: CI numbers, serial numbers, specification numbers, drawing or part numbers, change identification numbers, and code identification numbers. These identifiers, called configuration identifiers, are further described in Chapter 9. Using these identification numbers plus satisfactory compliance with the requirements listed below will provide the assurance of meeting the customer's identification and acceptance requirements to:

1. Identify contract documentation.
2. Identify drawings and engineering data.
3. Identify deliverable hardware and data (hardware identification may include bulk materials and piece parts).

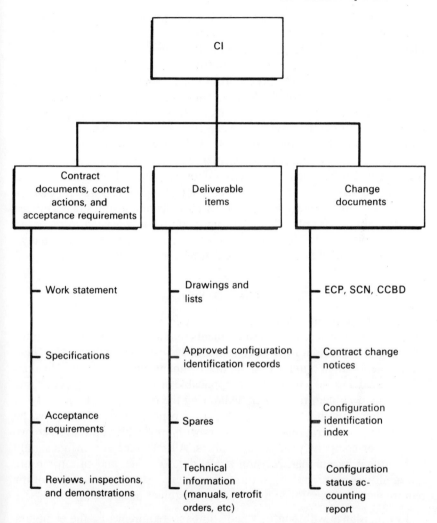

Figure 8.1 **Principal areas of identification for product and data acceptance by the customer.**

4. Identify technical information (manuals, special instructions, and retrofit orders).

5. Identify change documentation.

6. Comply with customer acceptance procedures and data needs.

7. Comply with review, inspection, and demonstration requirements.

The company is responsible for assigning identification numbers to each item given in the requirements listed above, using its own system and format, and for demonstrating that adequate policies, organization, and procedures exist to effectively implement the above identification and acceptance requirements. In particular, the company must have the capability to (a) identify and interrelate engineering data; (b) control and account for changes to data and the product; and (c) control the material, manufacture, and quality of the product. The procedures for achieving these requirements are given in Chapters 9 through 19. The following sections of this chapter are devoted to describing the requirements for each key identification area.

Figure 8.1 shows the products, documents, and actions that are identified and controlled during the project. It is the responsibility of the project personnel to use the company's system and procedures to tie these elements together and to reconcile engineering data with the deliverable hardware. Company systems and procedures must control engineering and production items below the product level, such as parts, components, subassemblies, and assemblies. Usually official customer control is restricted to the levels shown in Figure 8.1. Therefore, the customer must rely on the company's ability to control events and actions that he does not directly regulate.

8.2 RESPONSIBILITIES OF THE CONFIGURATION MANAGER

The configuration manager is responsible for determining the exact identification and acceptance requirements that apply to his project and for selecting the configuration management system that will satisfy these requirements at minimum cost to the company and customer. He does this by studying the contract and applicable specifications, by classifying and listing each item to be identified during the project, and by identifying the requirements that apply to each item. To provide an overall view, he may prepare a matrix relating each item with applicable contract reference numbers or company procedure numbers. All reference documents should be identified by number, revision letter, date of issue, and title to ensure traceability of requirements to their exact sources should the validity of the requirements be questioned by project or customer personnel.

In accordance with contract and company requirements, the configuration manager has three primary tasks during the project: (a) to compose and assign all configuration identification numbers; (b) to apply these numbers to data and products; and (c) to maintain the relationships and continuity among these data and the product so that product development, qualification, production, and logistics support[1] activities match engineering requirements.

Although customer identification and acceptance requirements should have been evaluated and mutually agreed to during the proposal and

---

[1] Refers to repair, maintenance, training, and supply of spares for the equipment.

negotiations with the customer preceding the issuance of the contract, the configuration manager should discuss with the project manager any problems related to interpretation or disagreement with the contract requirements. Customer requirements that will result in changes to the existing company system or procedures which meet the intent of the contract should be changed to allow the existing system to be used. When this occurs, customer verbal agreement should be documented through contract change notices. If numerous exceptions to requirements are requested, a written request is prepared describing each exception and the reasons for it. It is essential that the configuration manager show how the requirements will be met using the existing company system or that the requirements do not achieve their intended purpose.

8.3 ˙CONTRACT DOCUMENTATION

To define contract requirements and to control work authorized by the contract, the company may prepare documents based on the contract issued by the customer. The contract is a company-customer agreement that defines the terms and conditions under which the project will be conducted, the tasks to be performed, and the products to be delivered. In addition, a delivery schedule is given and applicable specifications, standards, and other documents are included by reference.

The contract is the key document for defining the project and CI requirements and is the final authority for acceptance of all deliverable items, including data. Any changes desired by the customer or company must be mutually agreed to and confirmed by a formal change document signed by both the customer and company contract administrators. Reliance on verbal agreements to change contract requirements will likely result in unacceptable equipment or data, or both, since the customer's representative is authorized to accept an item only if it meets the written contract requirements or modifications thereto. In addition, bad feelings between the two parties may develop because of misinterpretations on what was said.

Work Statement

A company-prepared work statement is a document that describes or expands upon certain contract requirements. The work statement is identified with the contract number, contract date, customer's identification, project title, CI number and nomenclature, and the company's internal work authorization number. The main function of this document is to describe in detail each major task that must be completed during the project. Each task description is identified with the corresponding contract article and line item number to ensure traceability to the original requirement. Other requirements include identifying each equipment, test set, spare component, and technical manual to be delivered to the customer with the CI number and nomenclature of each product. (In small projects, only one type, model, series product is usually required.)

Specifications

Specifications are the key source documents for defining the configuration of the product or its subordinate items. Therefore, accurate and uniform identification of these documents is a major requirement for all projects. Company specifications (system, performance, CI detail, and engineering and logistics critical components) directly related to the equipment and referenced in the contract are identified with standard configuration identifiers. Customer-imposed military and federal specifications and company standards retain their original numbers and are not reidentified unless the specification is revised by the company for internal company or particular project use.

In addition to the identification number, the revision letter and release date are used to identify a specification. If the specification is replaced with a new one, the new specification title page should include the following entry below the new identification number: Supersedes SP 34XX1, March 1, 1969. Other identifiers that interrelate the specification to the product, contract, part, and manufacturer are the CI number and nomenclature, part number, and the company's code identification number. All identifiers are placed on the specification title page. The specification number and release date are also added on the upper right-hand corner of each subsequent page.

8.4   ENGINEERING DRAWINGS AND DATA

Engineering drawings, parts lists, hardware allocation lists, and part standards are required to identify the design application, allocation, and revision of equipments, subassemblies, components, and parts. The design application of a part refers to its use in a higher level assembly, which may be identified by its item part number, nomenclature, next higher assembly part number, and CI serial number effectivity. The allocation of a CI refers to the system that it is a part of or to its geographic location. The goals of these data are to (a) establish common and standard parts usage to determine purchase, distribution, and inventory requirements; (b) determine product deployment locations so that spares and supporting facilities and products can be obtained and prepared to maintain the product; (c) identify product composition to aid in system integration; and (d) relate changes in product configuration, logistics support, and system requirements to approved change documents.

Drawings

Drawings, along with specifications, are the primary source documents for all configuration requirements for production of deliverable items on a project. Drawings are identified by document numbers, revision letters, item nomenclature, and manufacturer code identification numbers. These identifiers are located in a rectangular box at the lower right-hand end of the drawing. Key rules for correct identification and interrelation of drawings and associated lists are the following:

Number	Rule
1.	Once used for an equipment accepted by the customer, the technical description of an item or higher level of assembly must be retained on the drawing or in release records.
2.	The company assigns all configuration identifiers without approval of the customer. (Chapter 9 describes standard configuration identifiers.)
3.	All standard configuration identifiers and nomenclature, except serial numbers for items below the CI assembly level, are shown on drawings or release records.
4.	The same drawings and standard configuration identification numbers used to build or change deliverable products are also used to build spares and to write manuals.
5.	The same drawings and standard configuration identification numbers are used to build additional quantities of the same product.
6.	All drawings must contain the company's code identification number.
7.	All production drawings that describe a product must be interrelated by drawing application data (see Figure 8.2). The data listed below may be added to the drawings or their release records.

a. Part number, the item identification created by the drawing (usually the same as the drawing number). Figure 8.2 illustrates three part number entries for a multidetail drawing.

b. Next assembly, the drawing number of the immediate next higher assembly into which the item shown on the drawing goes.

c. End item designator, the CI number into which the next assembly is installed.

d. Serial No., the serial number of CI in which the item is to be installed. (The serial number is not always recorded on the drawing. It may be entered in the release record, depending on contract requirements or company policy.)

e. Top assembly for CI, which does not contain next higher assembly application data but includes the entry "Used On" or "As Allocated" for reference to customer usage; for example, the name of the system which uses the product is recorded.

f. Revision block, contained in all drawings. This block includes the revision letter of the drawing, the change identification number (CIN), a brief description of the change, date of revision, and the person who authorized the revision.

g. Revision description, the description and revision block (see Figure 3.3) which identifies the CIN (EO, ECP, etc.) and relates it to the drawing revision letter. It includes Class I or Class II for the type of change, ECP number if an ECP was issued for a Class I change, and the EO number as appropriate. Other pertinent change data may also be included.

12345-3	14000	Power Supply 12-345	05
12345-2	12500	Power Supply 12-345	05
12345-1	12400	Power Supply 12-345	05
Part Number	Next Assembly	End-Item Designator	Serial Number
Drawing and Part Application			

Figure 8.2   Drawing and parts application block.

Application Data. All drawings must contain application data for interrelating the drawing with other items in an equipment. The application blocks on the drawing may appear as shown in Figure 8.2 or may consist of only two columns: "Next Assembly" and "Used On" as shown in Figure 3.3. The drawing number of the next higher assembly for the drawing is given in one column. The "Used On" column identifies the final installation of the item and its parent CI; for example, assume a printed circuit board (drawing no. 12345) goes into a tuner chassis (assembly) (drawing no. 12400) which is then combined with other assemblies to form a radio, Model 200. The application data recorded in drawing 12345 would be as follows: "12400" would be recorded in the "Next Assembly" block and "Radio, Model 200" would be recorded in the "Used On" block.

Application data entries for top assembly drawings use different entries from other drawing levels. For a top assembly drawing the "Next Assembly" entry is usually left blank and the "Used On" block contains the subsystem or system name and model number that will contain the equipment described by the top assembly drawing.

Master Artwork. Artwork is used to fabricate printed circuit boards for electronic assemblies. Because of the dimensional accuracy required to build printed circuit boards, the original board layout is drawn on a large size polyester sheet. This document is then photographically reduced to a negative which is the actual size of the printed circuit board (for example, 2 by 3 inches). The original artwork is then stored for future use should the negative be lost or damaged or should changes be required. Since the artwork is really tooling and not a normal part of the drawing package, it should be identified and processed separately from drawings and parts lists. The artwork is usually identified with the same number as the circuit board detail drawing. By prefixing this number with MA for Master Artwork, the artwork can be uniquely identified and handled as an entity just as a parts list is handled. Thus the detail drawing and artwork could be identified as follows:

Detail Printed Circuit Board Drawing: 123455

Master Artwork: MA123455

Redrawn Drawings. A redrawn drawing retains its original identification number. However, the revision letter is increased one step and the revision description block receives this entry: "Redrawn with no change." If changes have been made, this statement changes to: "Redrawn with changes." If a change is made when the drawing is redrawn, the revision description and change authorization document number (such as EO 2010) are recorded. The old drawing is identified in the revision block as follows: "Replaced with (or without) changes by Revision B." The word "Superseded" is placed in letters larger than 1/4 inch above or near the title block, and the drawing is placed in the project history file for future reference.

Parts Lists

A parts list is a tabulation of all parts and bulk materials for building an item. When a parts list is an integral part of the drawing, the drawing identifiers also apply to the parts list and no special identification is required. If a separate parts list is used for the drawing, the list is identified with the same identifiers as are on the drawing except that a prefix PL precedes the drawing number to distinguish between the two documents. Parts and materials used to produce the item depicted on the drawing are listed along with the quantities and drawing reference designations. Each standard item or material in the list is identified by its industry, federal, or military standard number.

Nonstandard vendor parts or items are normally identified by vendor item identification and part number. The identification number of the specification or specification control drawing controlling the configuration of the vendor part can be used until the first production article is accepted. Only one vendor source can be listed for the product after this point unless additional sources are approved by the customer. If only one source is authorized, the component is identified by the vendor item identification part number. If alternate sources are approved, the component is identified by the specification identification number. (Under some contracts, the use of specification identification numbers for vendor item numbers is forbidden after FACI.)

When a source control drawing is required for a vendor item, the drawing identification number is used as the item number and is recorded in the "item identification" column of the parts list.

Hardware Allocation Document

A hardware allocation document is prepared for programs to identify the collection of CI's that constitute the system. The document is identified by a system designation number assigned by the customer and is entitled "Hardware Allocation Document." The purpose of the hardware allocation document is to identify the system configuration at each location, show the general arrangement drawings (top assembly drawings) of the equipments at these locations, and determine the location of system CI's by serial number. The document is prepared, maintained, and released monthly by the customer or contractor responsible for integrating the equipments into a system. Each company, however, is responsible for providing the integrating contractor with correct data on equipment configuration, location, and identification numbers.

Computer Programs

Computer programs required by the customer for the equipment are identified as end products or configuration items. Item identification is also required for computer program products such as card decks, discs, and magnetic tapes. These products are treated as end items or components and not as technical data; for example, the company may be required to identify each card by a part number, addressed to a next assembly deck(s) and to serial numbers of equipments that use the assembly deck. The same is true for each disc or reel of punched or magnetic tape. These items are identified by the item identification and part number, addressed to a next assembly deck and to the serial numbers of products that use them. A detailed discussion of operational computer software change control procedures is presented in Chapter 21.

Standard Parts

A standard part is an item that is subjected to more stringent and consistent manufacturing controls than a commercial item, which is available as an off-the-shelf item. An item is identified as a standard part if it falls into one of the following categories: (a) it is stocked and identified by federal stock number; (b) it is specified and described by a military standard; or (c) it is established as a standard and specified by a company source control drawing or other specification. Standard parts are identified by the company in accordance with contract requirements or company policies. If it is identified by the company, the identifier format may differ from the identifiers used for the project. Once a standard part has been referenced on a production drawing as a part of the product, it cannot be reidentified unless so directed by the customer. Because of its unique nature, a standard part should be identified so that it can be distinguished from other parts. This may require identifying only standard parts with a prefix ST or similar designation.

Top Drawings

In configuration management, the top drawing defines the configuration in which the product is to be removed and replaced for maintenance and modification; it also identifies the key assemblies that make up the equipment. The identification rules for drawings given in section 8.4 also apply to top drawings. However, as mentioned before, the "Used On" application block contains the system into which the product will go; for example, the solar wind spectrometer (product) is used on the ALSEP (*Apollo* Lunar Surface Experiments Package).

8.5 ITEM IDENTIFICATION

Deliverable products, spares, and other major hardware items are identified by nameplates and markings that are permanently attached to the item. All configuration identification data on these nameplates and markings are taken from and therefore agree with production orders, instructions, and release records.

Nameplates can be arranged as shown in Figure 8.3, containing the data shown. These nameplates are located on the product for convenient examination by service or operating personnel when the product is in its installed position. Engineering and logistic critical components are also identified as shown in Figure 8.3 except that (a) the family designation number is substituted for the CI number; (b) the part serial number replaces the CI serial number; (c) the component part number replaces the CI part number; and (d) the specification number and revision letter of the critical component specification replace the CI specification number and revision letter.

As already mentioned, punched cards, decks, computer tapes, and trays for deliverable computer programs are managed, accepted, and delivered as a manufactured product, not as engineering data. Therefore, each card is marked with its part number and the design activity code identification number. The band or case of each deck is marked with the complete item identification and part number of the deck and the company code identification number. The outside end of each computer tape is marked or punched for direct reading of the complete item identification and part number and the company code number. The tape reel and canister are also marked with the same identifiers.

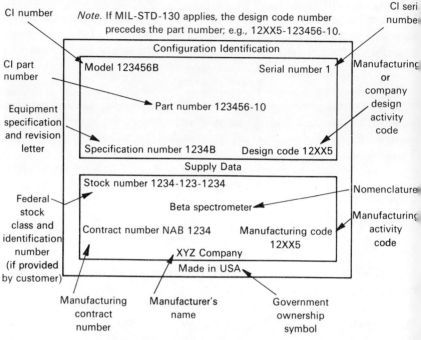

Figure 8.3   Sample equipment nameplate.

Products and critical components that are too small for complete identification are marked with the following minimum configuration data:

1. Part number.
2. Serial number.
3. Code identification number.

Serialization

The company is responsible for serializing and marking all products, engineering critical components, and logistic critical components. Other parts and subassemblies are serialized at the company's option. For best control of the configuration and for obtaining better traceability, serial numbers should be used whenever possible. However, if the additional costs are not warranted, serial numbers should not be used. The final decision should be made by the project manager after discussing the benefits and disadvantages of piece part serializing with the configuration manager. The guiding criterion for part, component, or assembly serialization is that if particular test data and test data sheets are required for record and acceptance, it follows that serialization is required to correlate the specific test data results. For critical applications, such as manned spacecraft, all items except bulk materials should be serialized and bulk materials should be identified by lot numbers.

Parts and Standards

All parts and company standards are marked or stamped with their item nomenclature (resistor, relay, analog-to-digital converter, and so on), part number or standard specification identification number, serial or lot number (if used), the code identification number, and manufacturer's name. Direct part marking is not required when the part is too small or would be damaged by making this identification, the part is encapsulated within another, or it is permanently assembled and not replaceable as a unit. When the part is not identified on its body, the shipping container should enclose the required identification data. Upon receipt of unmarked parts, they should be individually placed in plastic bags or boxes containing cards or slips of paper with the identification data.

Shipping and Storage Containers

Shipping and storage containers are identified with the standard configuration identifiers shown in Figure 8.3 or as specified by the contract. These identifiers must be permanently attached or inscribed on the containers in readily visible areas. Identifiers may be stenciled or printed directly on the containers or on gummed labels coated with lacquer. In addition to these identifiers, handling notes should be included. For instance, a space instrument container would be marked: "For Space Flight Use—Handle with Extreme Caution." This marking should be large enough to be read at least three feet away from the container. The quality assurance engineer is responsible for assuring that the markings are correct and complete before shipment to the customer.

Spares

Spare parts or components are identified as shown in Figure 8.3 with the CI number replaced with a spare part serial number. (Refer to Chapter 11 for a spares numbering system.) Note that spare parts are not serialized with the same format as the parts used in the product. The exact content and method of marking is usually specified by the customer. When the parts are too small for complete identification, then the project manager must select the identifiers after discussion with the configuration manager and the customer. As described above, small parts may be identified with tags containing the required data. The quality assurance engineer must inspect the markings before shipment to the customer.

8.6 TECHNICAL INFORMATION DOCUMENTS

A technical information document (also referred to as a technical order) is used to transmit technical data, instructions, and safety procedures related to the operation, installation, maintenance, and modification of a product. Correct identification of these documents is as essential as for drawings because these documents must be related to the proper product to prevent malfunctions or damage resulting from erroneous data. The functions of technical information identifiers are the following:

1. To relate the technical information to engineering data (drawings or specifications).

2. To identify the specific product for which the data are applicable.

3. To cross-reference the change, revision, or issuance of the technical information document to the approved engineering change being incorporated.

Technical Manuals

As described in Chapter 3, technical manuals provide detailed information on the operation, installation, or maintenance of a product. Therefore, when different product configurations are delivered to the customer, the manual is organized and identified so that it can be used to operate and maintain the different configurations. These multiple configuration descriptions are maintained until all products have been reworked to a common configuration. If the products are not reworked to a single configuration and differences among equipments are extensive, separate manuals may be necessary for each configuration.

The title page and a configuration chart are used to identify the manual. The first page includes the manual title and identification number, the CI nomenclature and number, CI specification identification number, part number if applicable, code identification number, manufacturer's name and address, and the date of issue. The configuration chart shown in Figure 8.4 relates the manual, its revisions, and supplements to the applicable configurations. The chart is inserted immediately behind the title page and ahead of a list of revised pages.

Time Compliance Technical Order

A time compliance technical order (TCTO) provides instructions for modifying a product, performing or initially establishing special inspections, or imposing temporary flight or operating restrictions. As indicated by the title of this section, these actions must be completed within a specified period. TCTO's describing retrofit actions resulting from Class I engineering changes (ECP's) are identified for configuration control and accounting at the CI management level.

Configuration Chart			
Technical Order (number and date)	ECP	Effectivity	Time Compliance Technical Order

Figure 8.4   Technical manual configuration chart.

In addition to the TCTO title, the CI number is added to the title page. The serial number of the equipment to be retrofitted is also recorded. To trace change incorporation into the field hardware and spares, company kit identification numbers must be assigned. Other identifiers include the item federal stock number and part number within the CI affected. Note that a TCTO applies to only one equipment serial number. Upon completion of the retrofit, the order is signed, dated, and copies returned to the company and customer as a record of retrofit accomplishment.

8.7   CHANGE DOCUMENTATION IDENTIFICATION

Change documentation to contract, technical data, and equipment configuration must be identified and interrelated for effective change control and accounting. In particular, the following change documents must be identified and maintained: engineering change proposals, contractual change notices, engineering orders, configuration identification indexes, and configuration status accounting reports. These last two documents are discussed in Chapter 19. All change documents should include as a minimum the following identifiers:

1. CI nomenclature and number.
2. Applicable serial numbers.
3. Company name and code identification.
4. Change identification number and date.

8.8   PRODUCT ACCEPTANCE

Upon satisfactory completion of the product and other deliverable items (data package, manuals, computer tapes, and so on), the company notifies the customer that the items are ready for final acceptance. Formal acceptance by the customer includes a review of all procedures, processes, and data necessary for building, inspecting, testing, and approving the product as well as data requiring transfer of ownership to the customer. Acceptance of an item by the government usually requires completion of a DD Form 250, Material Inspection and Receiving Report. This report identifies all delivered items by name and identification or part numbers. Each item is referenced to the applicable contract line item or paragraph entry in the contract. The DD Form 250 is signed by the government's quality assurance representative, indicating that the items listed meet contract requirements. In addition, the government representative authorized to accept and receive the items, for example, the project technical officer, signs the form. When all signatures are complete, the form authorizes the government's financial division to make a payment to the company (depending on the initial payment agreement made during the contract negotiations). If exceptions or shortages exist, these are listed on the DD Form 250 or its continuation sheet. If these shortages are deemed very important by the government, a percentage of final payment may be withheld until shortages or deficiencies are corrected.

Acceptance requirements can be simple or complex, depending on the nature of the product being accepted and its intended use. Several guidelines for acceptance are listed:

1. *Acceptance testing* to an approved test procedure is performed to verify that the product meets the functional and environmental requirements of Part II of the product or equipment specification.

2. *Identification* is verified by inspection of nameplates, markings, and title pages of manuals to assure correspondence with drawings prepared.

3. *Certification* indicates that the customer verifies that the product was built to the configuration required by engineering drawings, specifications, special processes, and customer-approved engineering changes.

4. *Specification* is reviewed to ensure that all waivers and deviations from this document are approved and accepted by the customer.

5. *Qualification status* refers to verification that engineering critical components are certified. A product is not certified as qualified if any one engineering critical component in the product is not qualified.

6. *Shortages* are recorded on the acceptance shortage record after a shortage review is made.

7. *Technical orders,* manuals and other technical information are verified for correct identification and instructions.

8. *Engineering data verification* indicates that verification of the required data is made to ensure that the specification, a set of released engineering drawings, the interface drawing, and the allocation document (see Chapter 8.4) are complete and meet contract requirements.

8.9 FORMAL REVIEWS, INSPECTIONS, AND DEMONSTRATIONS

Product reviews, inspections, and demonstrations are performed as required by contract. These reviews are held with customer representatives at key milestones during the project and generally result in identification of baselines for control of changes to the configuration of the product or in the acceptance of production equipment. Note that although the customer officially participates in these reviews and gives his approval of the product configuration, the company is still responsible for meeting all contract requirements. The reviews, described in Chapter 14, include the following:

1. Preliminary design review (PDR)/system functional audit (SFA) (optional).
2. Critical design review (CDR)/functional configuration audit (FCA).
3. First article configuration inspection (FACI)/physical configuration audit (PCA).
4. Final configuration review (FCR).
5. Flight readiness review (FRR).

The last two reviews (FCR and FRR) are usually held at the operational site under the direction of the system manager.

In parallel to the contractual reviews described in the preceding discussion, the company implements an internal design review program carried out in appropriate stages (see Figure 14.1) to cover such areas as systems concept, design philosophy, trade-offs, part applications, materials, circuits, processes, structures, and thermal design.

There are four checkpoints where design reviews may be applied:

1. Conceptual design review, to be done as early as possible after basic concepts have been defined.
2. Preliminary design review for electrical and electronic products, accomplished after circuit design and breadboard testing are complete. For mechanical, structural, and propulsion products, to be done after initial design and engineering model testing have been completed.
3. Development design review, immediately before qualification model fabrication and test.
4. Preproduction design review, done after all drawings are complete.

These design reviews are discussed further in Chapter 14.

Chapter 9

CONFIGURATION IDENTIFIERS

The importance of numbers for identifying the configuration of an equipment has increased enormously with the advent of the configuration management discipline. As a result of this new discipline, formal requirements have evolved for most contracts with the government that require the company to be capable of relating and tracing contract, administrative, and management actions to equipment descriptions. In addition, these descriptions must be completely and uniquely identified by six standard types of configuration identification numbers that are the foundation for effective configuration control and accounting. These numbers provide complete technical and contractual identification of the equipment configuration and permit connecting its design to contract and administrative actions controlling development, fabrication, and testing activities.

The six standard configuration identifiers used by the government and industry for complete identification of end items, such as hardware and engineering data, are the following:

1. Specification numbers.
2. Equipment numbers.
3. Drawing and part numbers.
4. Equipment and item serialization numbers.
5. Change identification numbers.
6. Manufacturer's code identification numbers.

A review of these identifiers indicates that they provide complete identification and control of key data and hardware and enable the company or customer to obtain the following information on a CI or a lower level item:

1. Technical requirements for the equipment that provided the basis for detail design and construction.

2. Identification of the equipment designed and built to the applicable specification.

3. Technical descriptions for the equipment and its lower level items.

4. Description of the sequence of manufacture of the equipment, its parts, test data, and the engineering changes made to each equipment by serial number identification.

5. Change documents affecting the equipment and their authorization for implementation.

6. Identification of sources of manufacture at all levels of the equipment.

As can be seen, the requirement for a complete knowledge of the equipment's configuration is met by the above capabilities. Since the standard configuration identifiers are the only officially approved identifiers used to obtain information on the CI configuration, their issuance and control must be strictly regulated by the company. Failure to regulate the release and revision of these identifiers in accordance with technical or administrative changes can result in the company's inability to accurately identify the configuration of each equipment delivered to the customer and may result in substandard performance, poor operational support, injury to personnel, or damage to the equipment; for example, assume that a power supply module used in the first equipment delivered to the customer has a design change made that makes subsequent modules noninterchangeable with the module used in the first equipment. If the company merely assigns a change letter A to the module drawing number, the configuration difference between the first and all subsequent equipments will be lost because the part number is essentially unchanged (when a part becomes noninterchangeable with previously built ones, its number must be changed). As a consequence, during field use replacement of a defective power supply in the first equipment will be made with a power supply built to the revision A. Assuming that pin 9 is grounded on the first power supply but has 200 volts on subsequent supplies, the result of placing the new supply into the first equipment will cause a short circuit that will probably burn out a part or component unless the protective devices open fast enough to protect the circuit.

Configuration identifiers, groupings of numbers and letters, are assigned to an equipment and its documentation by the configuration manager or his representative. The configuration manager is responsible for assigning, controlling, and verifying that these numbers are being used correctly according to the applicable contract documents or company policies that give the system for identification numbering. A key area for him to watch is that of changes to configuration identifiers on items provided by other sources. When the equipment incorporates the design of the customer, subcontractors, vendors, or suppliers, the company is required to use the configuration identification numbers assigned by these groups, unless exceptions are authorized by the customer or the governing document for equipment identification.

A summary of the characteristics, use, and relationships of standard configuration identification numbers is given below. Table IV shows samples of these numbers. The detailed numbering systems and procedures for the various types of numbers or identifiers are given in the following four chapters (10 through 13), which cover specification, equipment, drawing and part, serial, and change identification numbers.

TABLE IV   Standard Configuration Identifier Formats

Identifier Type	Sample Identifier Formats[a]
Specification	CP 444100A
Equipment	CI 123456A[b]
Drawing	123456C
Part number	123456-1
Serial number	SN 4, SN 04, or SN 004
Change identification	
Specification change notice	5-20b
Engineering change proposal	445200C 1112 SWS 606 15-1 D R2-C1
Engineering order	EO 1234
Code identification	09876

[a] Significance of letters and numbers is described in the following chapters.

[b] The letter "A" used in this case is not a revision letter but a model change designator and has no connection with the drawing revision letter.

9.1   SPECIFICATION NUMBERS

The specification number is assigned to the document that identifies the technical requirements for the equipment. The specification number identifies equipment and component specifications and ties these engineering documents to the project specification tree and appropriate equipment baseline. Specification numbers are also used to identify company standards and military specifications or standards applicable to the project. Military standards are used frequently in the aerospace industry to obtain reliable parts conveniently. These parts are ordered in conformance to the applicable military standard document, which imposes stringent design, construction, and quality control requirements for the parts.

Note that capital letters are added after the number to denote a revision to the specification. These letters must be included in the application or referencing of the specification. Specifications referenced in drawings and parts lists, however, are an exception. The specification revision letters are omitted unless the exact issue of the specification is to be used. This is done to avoid changing the reference on the PL each time a specification is revised.

Chapter 10 contains a detailed description of specification numbers.

9.2   EQUIPMENT NUMBERS

The equipment (CI) number identifies the type, model, series required by the customer. A type, model, series is any quantity of equipments of one basic design that is defined by one detail or governing specification as a block of items to be designed, developed, and built to that specification. It is a permanent number assigned to the equipment and identifies the level at which technical and contractual requirements are controlled; for example, all requirements or references are tied back to this number. The equipment number shown in Table IV has an "A" at the end of it. This is not a revision letter. It indicates that the equipment is the second in a series of equipments procured by a customer. Although the equipments in all the series are similar, each series of equipments has some design features that are different from the preceding or following series. Chapter 11 discusses equipment numbers in greater detail.

9.3   DRAWING AND PART NUMBERS

A drawing number identifies the document that describes an item or the equipment. This number is assigned by the company's design group and is usually the same as the part number or is integral to the part number. In other words, if the part number is different from the drawing number, the drawing number must be included in the part number. A key feature of a drawing and part number is that it uniquely identifies the equipment because an identical drawing and part number is never released or issued for more than one item. These numbers are also used to control the assembly and replacement of parts, components, subassemblies assemblies, and the equipment itself. Refer to Chapter 12 for a detailed description of drawing and part numbers.

9.4   SERIAL NUMBERS

Serial numbers provide a means of controlling configuration effectivity (see Chapter 7.6), facilitating traceability and accounting of individual components and assemblies, and their installation and location within the product. A serial number is used to differentiate between two or more parts or assemblies within the same part number group (family designation number).[1] Serial numbers, once assigned within a part number group, never change and are never reassigned. If an item goes out of existence, the serial number is taken out of circulation. If an item is reworked, the serial number is assumed by the superseding item. Chapter 11 provides additional information on serial numbers.

9.5   CHANGE IDENTIFICATION NUMBERS

The change identification number is assigned by the configuration manager to identify an engineering change subsequent to its initiation and is retained throughout its entire processing cycle. Once assigned, engineering change identification numbers may not be reassigned, even though the change involved is disapproved. All engineering data, directives, planning, manufacturing and purchase orders, and other pertinent documentation generated by an approved change are identified by the change identification number assigned. The change identification number can be of many versions; however, a uniform company-wide system is essential. For example, the number may be assigned from a sequential nonsignificant[2] block of numbers within each project. Some projects may choose to apply significant number schemes into the change number which distinguishes individual CI applications, subsystem, or product-line effects. The only mandatory requirement laid down by the government relative to significant numbering is with respect to "change packaging." In many cases, a change to one CI will necessitate a change to others. The change identification number must facilitate a common denominator within the change number composition which correlates and groups all CI's affected. This is usually accomplished by dash numbers.

Configuration Item (or product element)	Change Identification Number
Radio (receiver)	735
Transmitter	735-1
Test set	735-2

The ECP number is often used to identify a change during its processing cycle. Chapter 13 discusses the structure of engineering change numbers in greater detail.

[1] Part number group (FDN) is defined as an original item and all subsequent items of the same basic number designed to replace or supersede it.

[2] Numbers which do not contain coded information about the part.

9.6   CODE IDENTIFICATION NUMBERS

The code identification number identifies the design or manufacturing source of the equipment. This is a government assigned number used for uniquely identifying every company or agency that builds or designs hardware for the government. Companies, such as dealers and vendors, that neither build items nor control design for the government are not issued these numbers.

The code identification number is recorded in the title block of the document as shown in Figures 3.3 and 3.4. Usually the number is preprinted on drawing vellums with the title block, margins, and company name. Used with the preceding identification numbers, it guarantees a unique identification of all engineering documents. It is a five-digit number (e.g., 54XX1) of the company as shown in *Military Handbook* H4-1, "Federal Supply Code for Manufacturers, Name to Code." Large companies having several divisions may have separate codes assigned for each division.

The federal supply code[3] is obtained by writing to the Defense Supply Agency, Defense Logistics Services Center, Attention DLSC-CGC, Federal Center, Battle Creek, Michigan 49016. Your request should include the government projects that are in progress at your company and their contract numbers.

The use of the code identification number in parts lists is not covered in the following chapters and is therefore presented here. When a part is identified on a parts list, it is usually preceded by its manufacturer's code identification number. However, three conditions may exist which eliminate the need for recording the identification number with the part number.

1. The part is a standard or specification item which is identified in an industry association standard or in the "Department of Defense Index of Specifications and Standards" (DODISS).

2. The contractor, generating the drawing or parts list, need not list his code number against each of his own part numbers; that is,

   a. The code number is the same as that shown in the title block of the drawing.

   b. The code number is the same as that shown in the title block of the parts list.

9.7   ADDITIONAL IDENTIFIERS

Numbers other than the six standard identification numbers described above are not as directly used for configuration management. However, other

---

[3] Same as code identification number.

identifiers, used for internal company functions, may be considered supplemental to configuration identification requirements. These numbers include: (a) customer registration numbers; (b) equipment type or class designations assigned by the customer; (c) federal stock number (FSN) assigned by the government for inventory control; (d) manufacturing numbers, other than serial numbers, that denote special data about items being manufactured; (e) synthetic part numbers used by the company to identify a subassembly not shown in a separate engineering drawing; and (f) material codes for material control by the company. An example of a material code is the lot number C1002 assigned to an item or group of items received from a vendor. The C identifies the third project or contract received by the company and the 1002 indicates that it is the one thousand and second parts shipment received for project C. This lot number, assigned by the material controller or quality assurance engineer, provides a unique company identifier that enables the part to be traced to its original receiving report, which includes all the data required to identify its source, lot or serial number, certification data, etc. A log book is maintained to cross-reference the lot number to the purchase order and receiving report.[4]

9.8 STANDARD CONFIGURATION IDENTIFIERS

The standard configuration identifiers described in sections 9.1 through 9.5 are discussed in fuller detail in Chapters 10 through 13. Keep in mind that the numbering systems presented are not sacred and may be modified by each company to meet its specific needs at lowest cost and labor requirements. Of course, when government contracts impose identifier contents and formats, these must be followed. If government requirements are presented as guidelines, the company may follow its own numbering system, modifying where necessary to meet the intent of the government's requirements.

---

[4] A receiving report identifies the items received in a shipment from a vendor. It includes quantities received, part numbers, serial numbers, lot numbers, and the number of the original purchase number.

Chapter 10

SPECIFICATION IDENTIFICATION NUMBERS

All specification documents must be uniquely identified before release for use. Because of their critical nature to a project, extreme care is required to prevent errors in their classification and identification. Specifications should be classified in accordance with their function or end use; for example, a specification can be used for procurement of an item from a vendor, for defining a product's performance requirements, or for defining a CI's detail design and construction characteristics. The specification identifiers should be formatted to allow for easy and positive differentiation among the various types of specifications. This can be done by using significantly different formats or by using a different prefix letter code for each type of specification. Specification identification within each type of specification is obtained by assigning different numbers to each document prepared.

The exact format and length of the specification number is not as important as company-wide consistency in its use. However, simplicity and brevity of identification numbers provide for easier retention, identification, and recording. Also fewer errors will occur in transcribing the numbers to various records and documents used on the project.

Specifications must be reidentified whenever they are changed. Usually, a revision letter is added to the end of the identification number and a new issue date is included at the right-hand top of the title page. In other cases, a completely new number may be assigned. A central authority should process all changes to specifications and issue the new documents.

The configuration manager and data control personnel are responsible for assigning and controlling identification numbers for specifications, specification change notices, and specification revisions. These numbers apply to all specifications and standards required to control the design and construction of products to be accepted by the customer. Specification number composition, assignment, and changes are described in the following sections.

10.1 COMPOSITION OF SPECIFICATION NUMBERS

Design and process specifications, specification control drawings, and source control drawings are identified by a specification identification number, which may include as many as 15 alphanumeric characters. This number usually consists of a prefix code, an identification number, and a suffix code as follows: PS 1234554B. In many companies specification and source control drawings are identified with regular drawing numbers or with drawing numbers reserved especially for these kinds of documents; for example, only drawing numbers beginning with 90 are assigned to specification or source control drawings (see Glossary), or an ST or C prefix may precede the drawing number.

Specification change notices are identified by a number consisting of an SCN number, a dash, the affected specification page number, and a suffix letter. The basic format for an SCN number is as follows: 5-20b.

The identification numbers of government or industry specifications and standards are used as specified, except when the specification or standard is changed by the company and reissued as a new document. Then a new number is assigned to the document by the company.

10.2 ASSIGNMENT OF IDENTIFICATION NUMBERS TO PRODUCT SPECIFICATIONS

The configuration manager assigns the product specification a number which consists of a prefix, document number, and suffix, unless a number is already assigned by the customer. Once assigned to a customer approved specification, the number cannot be changed or reassigned to another specification because of the possibility of having two different specifications identified with the same number. Table V shows sample specification numbers. The purpose of Table V is not to prescribe format style or number composition, but to illustrate one approach to specification identification number assignments.

The prefix is a set of two capital letters that identify the type of specification. For example, during initial release of NPC 500-1, the National Aeronautics and Space Administration (NASA) required the following prefixes for eight types of specifications:

Prefix	Specification Type
PS	Performance
RS	Project
SS	System
CP	Prime equipment
CF	Facility
EC	Engineering critical component
LC	Logistic critical component[1]
ST	Company standard

The use of these prefixes is no longer required because of the difficulties involved in having industry change its existing document identification numbering systems. Current NASA and DOD (MIL-STD-490) requirements permit companies to identify specifications with their own numbering system as long as the code is uniformly applied.

One final comment on addendum specifications. An addendum specification is a specification for a new equipment, but it is prepared from a specification that was used to make an existing product. This specification includes by direct reference applicable paragraphs in the original specification. When an addendum specification is written from an existing specification, the configuration manager assigns a completely new number to the document as shown in Table V. This addendum specification is maintained by its own change description and revision sequence and is completely separate from the original document.

The suffix code consists of a capital letter assigned in alphabetical order beginning with the letter A; it identifies the latest approved specification revision. The letters I, O, Q, S, and X are not used for revision letters because

TABLE V
Sample Specification Identification Numbers

Specification Number	Description
CP 123456	Initial release of prime equipment specification.
CP 123456A	First approved revision to specification.
CP 123456B	Second approved revision to specification.
CP XXXXX	First approved addendum specification made from CP 123456 and to be maintained by SCN's and revisions independent of the specification. (It has an entirely new number construction.)
EC 123456	Component specification maintained by company.

[1] An LC specification applies to repairable items that require the customer to buy spares or that are to be purchased from more than one vendor.

they can be confused with similar looking numbers or letters; for example, O and Q or 5 and S. The revision letter is shown below as a part of a specification number.

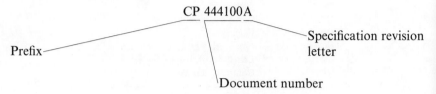

CP 444100A

Prefix

Specification revision letter

Document number

10.3   ASSIGNMENT OF NUMBERS TO SCN's

A specification change notice (SCN) is a record of changes to an approved specification that is attached to the specification when approved by the customer. SCN numbers are assigned by the configuration manager for each SCN to be prepared. One SCN is required for each specification change approved by one engineering change proposal (see Chapter 5.6). The SCN number identifies the SCN as a part of the specification that it changes, and it is formatted as follows: the first character is a numeral assigned in numerical sequence beginning with the number 1 for the first SCN prepared. The SCN sequence number is a common identifier for all SCN sheets prepared as a part of one specification change. Once the SCN has been submitted to the customer for approval, its sequence number cannot be changed or assigned to another SCN within the same specification. Thus, rejection of the SCN by the customer does not release the SCN number for use elsewhere. However, if the SCN is cancelled before submittal to the customer, the SCN number may be reassigned or recorded and retained in the system as a cancelled SCN at the option of the configuration manager.

The SCN sequence number is followed by a dash and the page number of the specification affected by that sheet of the SCN. The content of each SCN sheet applies to one page of the specification; however, multiple sheets can be used when more than one sheet is required to include all the changes. The specification page number is followed by a lower case suffix letter in alphabetical sequence for each SCN sheet. This suffix is used only when more than one SCN sheet is required to describe the change for a single specification page. An example of an SCN number follows:

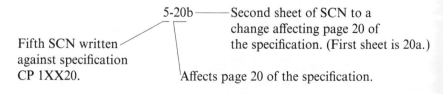

5-20b —— Second sheet of SCN to a change affecting page 20 of the specification. (First sheet is 20a.)

Fifth SCN written against specification CP 1XX20.

Affects page 20 of the specification.

The SCN number just given is meaningless unless it is keyed to a particular specification number. Therefore, any reference or record for SCN numbers must include the specification number and its revision letter.

10.4   CHANGES TO SPECIFICATION NUMBERS

The prefix and document identification numbers of a specification number are permanently assigned, unduplicated identifiers that uniquely identify the specification. Customer approved changes to a specification are identified by capital letters added to the end of the number as described in Chapter 10.2. Changes are identified sequentially beginning with A and increasing by one letter for each new approved change; for example, A for the first change, B for the second change, and C for the third change.

Amendments are brief, separately issued documents that correct, add to, delete, or clarify requirements given in the specification. Amendments are usually limited to a small number of changes. If extensive changes to the specification are required, the specification should be revised and reissued. Identification of amendments may be done as follows, with the information recorded in the upper right-hand corner of the title page:

> Specification No. CP 123456B
> Amendment 1
> January 3, 1970

Supplements, like amendments, are brief revision documents that are used to modify a specification and are issued as separate documents. The purpose of a supplement is to give special requirements for limited application(s). Supplements are identified as follows, with the information recorded in the upper right-hand corner of the title page:

> Specification No. CP 123456B
> Supplement-1B
> October 19, 1970
> Superseding Supplement-1A
> May 5, 1969

Note that a revision letter is assigned to a new issue of a supplement. In this case Supplement-1A was changed to 1B. In some companies, only one supplement may be allowed to be in effect at one time.

10.5   ASSIGNMENT OF NUMBERS TO COMPANY STANDARDS

The company assigns identifiers to each process specification, specification control drawing, source control drawing, and standard to be prepared and maintained by the company and identified on product drawings and specifications. A company standard gives common technical requirements for a part, material, process, or protective treatment. Care should be taken not to identify a company or commercial part as a standard when performance histories, tests, or the engineer's judgment provide doubt as to the parts' ability to perform consistently within the required performance and quality standards of the product built by the company.

The configuration manager, data control, or a standards group within the company is usually responsible for assigning identification numbers to standards. However, the configuration manager should monitor the identification numbers of standards selected for use on his project. In addition, he should make sure that the customer is notified in writing of any revisions or reidentification of standards specified on equipment drawings or parts lists. A copy of the latest revision should also be mailed to the customer. The company standard identification number may be assigned as follows:

Company standard prefix code is indicated by the first two characteristics of the identification number, the capital letters "ST" or some similar designator. These letters indicate that the number represents a company specification or standard.

Company standard identification number follows the prefix code shown in Table VI. This number includes codes to identify fixed (constant or common) characteristics of a related group of standards described by a tabulated specification or drawing. A tabulated drawing or specification

TABLE VI
Sample Company Standard Identifiers

Prefix	Number[a] (fixed characteristics)	Suffix[a] (variable characteristics)	Description
ST	7R6U	-6-12	Company standard with four coded fixed characteristics and two variable characteristics.
ST	8R6U	-6-12	Company standard reidentified as a new standard because of a characteristic change.
ST	8R6U	-6-12-3	Company standard reidentified as a new standard because a variable characteristic was added.

[a] Each letter and number identifies a characteristic of the item. A legend is included in the standard or other document for describing each characteristic.

shows similar items, which, as a group, have constant and variable characteristics. Each item is uniquely identified on the document. The advantage of a tabulated document is that it obviates the necessity of preparing an individual document for each item tabulated. The formats shown in Table VI are samples and may differ from company to company. For example, a standard may be identified by company Ace-Space (AS) with AS-STD-101. (STD equals standard.)

Company standard suffix codes follow the document identification number. These codes identify the variable characteristics of the related group of standards described by a tabulated specification or drawing. For example, the following characteristics could be described by the suffix codes: mounting dimensions, lengths, protective treatments, input and output voltages, and tolerances.

Specification Control and Source Drawing Numbers

Specification control and source control drawings are basically drawings (see Chapter 3) and therefore drawing numbers are assigned to these documents by the design and drafting group. Changes to these numbers are made as to drawing numbers, and not as to specifications described in this chapter. Refer to Chapter 12 for specification control drawing numbering and revisions. However, remember that specification control numbers are not part numbers and should not be used as such; but source control numbers are used as part numbers (see MIL-STD-100A, p. 58). When more than one vendor is listed for a repairable item and repair parts are not interchangeable, dash numbers should be used to identify each vendor item.

Company Standards Prepared from Other Existing Standards

Specifications and standards are assigned numbers by the government and industry. Once assigned, these numbers are not normally changed. However, new identifiers must be assigned when it is necessary to alter the existing standard specification to meet the product performance, reliability, and quality requirements and because the existing standard can no longer be used for this application. In producing the new standard, the company format, style, and identification system should be used.

Company Standard Revision Letters and Source Codes

In addition to company standard identification numbers, a source code (manufacturer's 5-digit code number) and a revision letter are used for identification of the document. However, reference to the specification identification number in a drawing or parts list should not include the revision letter unless a standard or specification identification with that particular revision letter is the only one to be used. The reason for this policy is to avoid constantly revising drawings and parts lists to include the latest revision letter because a listed specification or standard has been revised. When the standard is to be used during manufacturing or testing, the quality assurance and manufacturing engineers are responsible for ensuring that the latest revision document is obtained and used. The source code is usually placed on the title page with the identification number and revision letter. Each following page should also carry the identification number and revision letter.

Limitations of Company Standards

The specification identification number is an effective identifier only if the standard specification document is carefully written and defined to assure that part number changes in the internal construction of the standard, which cannot be identified and therefore controlled, will not adversely affect the function of the product. Qualifying and receiving inspection criteria should always be contained in the standard specification to the extent necessary to assure proper application and repeatable performance.

Identification of Standards in Drawings and Parts Lists

Standard parts can be identified in drawings and parts lists by government, industry, or company standard specification identification numbers rather than by their individual part numbers. The advantage of this system is that it reduces piece part inventories because parts with different numbers and vendors can be stocked in one bin, identified with the standard's identification number as long as all parts conform to the requirements of the standard. Without the standard, each part must be separately stocked and inventory controlled. In the electronics industry, where thousands of standard piece parts may be stocked, the storeroom size may be reduced by one third as a result of using standard or specification identification numbers. Of course, the vendor's parts numbers are always retained on the drawings or other related documents. Identification numbers for industry and military standards are usually prefixed with AN, MS, AS, or NAS designators. (See Acronyms for prefix descriptions.)

Reidentification of Standards

Standards are stable documents that do not normally require revision or reidentification. However, they do get changed and sometimes require reidentification with a new number. For minor changes not affecting the configuration or interchangeability of the part, only revision letters are necessary. When a significant design change is made affecting the configuration (performance, size, interface characteristics, and the like) the standard must be reidentified with a new number. Note that release of a new standard does not necessarily result in obsolescence of the original document. Whether the old document remains active depends on the nature of the changes made to it. Five reasons for reidentifying a standard are listed:

1. A new characteristic is added.
2. Changes in characteristics result in disqualification of one or more previously approved vendors.
3. Quality control requirements are made more severe.
4. Changes in the characteristics require rework or reidentification of products in production.
5. Changes in characteristics require retrofit of products delivered to the customer.

Standards may be revised without requiring issuance of a new number. As mentioned previously, minor record changes, such as correction of misspellings or incorrect grammar, do not require reidentification of the standard. Changes to improve clarity or to provide a fresh vellum when the old one is torn or damaged merely require a revision letter addition to the document number.

Another change not requiring reidentification of the standard is: additional tabulations of an existing variable characteristic are incorporated without changing previously tabulated variable characteristics. For example, the addition of two more tabulated voltage ratings for a transformer does not require reidentification of the transformer standard.

The addition or deletion of an approved vendor does not require reidentification of the standard if the fixed and variable characteristics specified in the standard are not altered and the quality control requirements are not the reason for deletion of a vendor.

Chapter 11

CONFIGURATION ITEM AND SERIALIZATION NUMBERS

Configuration item and serialization numbers are the two most important standard identifiers used for maintaining control over the product configuration. Therefore, the assignment and construction of these numbers should be carefully evaluated and established at the beginning of the project. Because of their importance and relationship to each other, they are discussed together in this chapter.

The configuration item number is permanently assigned by the configuration manager to identify all equipments comprising one family (type, model, series). This number refers to a family of equipments for a spacecraft mission, aircraft, ground vehicle, facility, or any other application. As an example, ten identical telemetry encoders to be used on a series of spacecrafts represent an equipment family. Whether one, two, or three encoders are used on each spacecraft does not change their family designation.

One method of assigning CI numbers is to automatically assign the first number in the block of drawing numbers issued to the project as the CI number. Since blocks of drawing numbers are carefully controlled (or should be), a unique equipment (CI) number is issued. It is *very practical* to have the CI number be (a) a portion of the specification number or (b) the top assembly drawing number.

Serial numbers assigned to products fall into three classes:

1. Customer serial numbers.
2. Configuration item serial numbers.
3. Traceability control serial numbers.

The product level at which each class of serialization is applied is established by the configuration manager prior to release of drawings.

Customer serial numbers are those serial numbers specifically assigned by the buyer. Customer serial numbers, when assigned, are correlated with the configuration item serial number and are shown in combination on the configuration item nameplate. However, this condition is not common.

Configuration item serial numbers are those serial numbers preassigned by manufacturing to individual CI's.

Traceability control serial numbers are used when traceability of an individual item is required. Special control serial numbers are sequentially assigned to the detailed parts, subassemblies, and assemblies. Typical examples of traceability control serialization levels are as follows:

1. Electronic Equipment is serialized to the lowest level below the configuration item at which test data is to be recorded; i.e., modules, assembled printed circuit board, special tested and "screened" components, etc.

2. Mechanical Equipment is serialized to the lowest level at which data is to be recorded; i.e., functional mechanisms and skeletal structures or substructures onto which other equipment is assembled.

The following sections cover detailed discussions of configuration item and serial numbers.

11.1 CONFIGURATION ITEM NUMBERS

The configuration item number identifies all equipments for a type, model, series. An equipment type, model, series constitutes a block of deliverable equipments that have the following characteristics:

1. The type, model, series of equipments is a requirement in the contract.

2. All equipments are designed and controlled by one detail equipment specification.

3. All equipments are identified and described by one top assembly drawing and a subordinate structure of installation, subassembly, and detail drawings.

4. Configuration differences between equipments are identified and described by adding or limiting the design application of the part or subassembly comprising the affected equipments.

5. Each equipment is formally accepted by the customer.

6. The type, model, series is used as the foundation for providing spares and for preparing operating and maintenance manuals for the equipment.

The equipment number is assigned by data control when the equipment specification number is assigned or when the production drawing for the equipment is released, whichever comes first. A typical equipment number is the following: CI 123456A. The equipment number may be (a) a portion of the specification identification number or (b) the top assembly drawing number for the equipment. The last character of the equipment number is a serial letter "A." Once assigned, this letter is the permanent series designation for all equipments for the type, model, series. (Do not confuse this letter with the revision letters of the top assembly drawing, which may be an A, B, C, etc.). Follow-on contracts for the same type, model, series retain the same letter. A new letter is assigned when a new series is specified by a new equipment specification; a new top drawing; or a new basis for acceptance, processing, operation, and maintenance is established.

11.2 SERIAL NUMBERS

Equipment serial numbers usually begin with the number 1, 01, 001, and so on, and are permanently assigned in numerical sequence within one equipment family or type, model, series. The CI number, assembly part numbers, and serial numbers define the engineering effectivity of the design. These numbers must be permanently marked on each equipment to preserve correct identification throughout its life. Serial numbers are also recorded on release records, EO's, ECP's, and configuration status records for relating engineering and management actions to the affected equipments.

The same serial number sequence is retained when the part number is changed to identify a noninterchangeable design or when the equipment is reworked or retrofitted. Thus, if a series of six voltmeters is being built, the meters will be serialized as 1, 2, 3, 4, 5, and 6, regardless of the engineering or fabrication changes that are made.

Serial numbers are placed on an item by manufacturing when the item assumes an identifiable configuration, such as at a major jig point or at the beginning of final assembly.

Spares are replacements for worn out or damaged items. When serial numbers are recorded on manufacturing drawings, serialization of spares is usually handled in a different manner from normal production items. Since the quantities and delivery schedules for spares are not firmly fixed when the drawings are prepared, a separate serialization format is used to simplify identification and traceability. This format consists of a sequence number for the spare, a dash, and the applicable production item serial number. For example, the serial number 15-201 represents the fifteenth spare built to the configuration that was in effect for production item serial number 201. Using this system, no new drawings have to be made indicating the serial number effectivity of the spares because the latest production drawing is available, which includes all the revisions made during the project.

The preceding serialization methods are not the only ones followed. Other schemes can be used to suit the company as long as they do not conflict with the customer's requirements. In some cases the customer may require a company building three photometers for use in a spacecraft to assign the serial numbers 301, 302, and 303 to the equipments. The last digit represents the photometer sequence of manufacture and the two first digits (30) could be used to identify the next higher assembly into which the photometer goes.

Another approach could be to assign serial numbers to an equipment which identify its class (prototype, flight, or spare) and the program that it is made for. Following are three sample formats for the *Nimbus* spacecraft program:

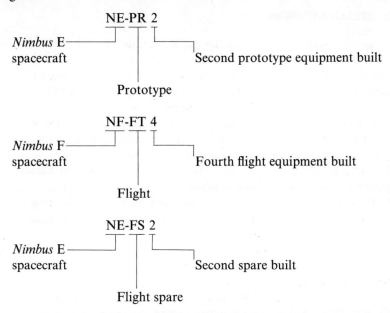

Chapter 12

DRAWING AND PART IDENTIFICATION NUMBERS

Drawing and part identification numbers represent the largest quantity of standard configuration identifiers related to a CI or project. For small projects, a few hundred drawing numbers may be issued and individual part numbers may exceed 1500. To avoid loss of configuration control, it is imperative that drawing and part number format selection, composition, and assignment be accurately and carefully performed.

The format and length of company-assigned drawing and part numbers are determined by the engineering and management groups. An unbelievable variety of formats exist in industry and their content depends on the objectives and needs of the originating groups. However, as time passes new personnel replace the originators of the existing system and in many cases object to the numbering system selected by their predecessors because it is cumbersome, inflexible, lacks coding information, or is not used as intended. Consequently the temptation to establish a new, improved system is great. Before any action is taken, management should make a thorough evaluation of both the existing and the proposed identification systems. If the benefits of the new system are not overwhelming, the old system should be retained to avoid the horrendous task of reidentifying drawing and part numbers for projects currently in progress. Even if current projects are excepted, conversion to the new numbering system is a formidable task requiring revision of procedures and records and reeducation of personnel. Of course, if a significant cost reduction can be demonstrated, management should encourage the change.

As mentioned before, identifier formats can take numerous shapes and lengths and vary so drastically in industry that in most cases the identifiers used by one company have no relation to those used by another company. In fact, the government maintains a catalog of company numbering systems and requests companies to provide descriptions of their systems for this catalog.[1] Although greater conformity is desirable, this situation is acceptable if the identification number formats satisfy the varying objectives of the companies using them. A serious problem exists, however, when divisions within one company have different systems. For example, if Division A and Division B of a company have their own systems of identification, the economical advantages of common requirements for parts or drawings are lost, as in the case of an analog-to-digital converter module that is used by both divisions and is purchased from the same vendor. In this case Divisions A and B order the same part separately, thereby losing the lower unit prices that would result from their larger combined order, as well as duplicating purchasing, inspection, and inventory control operations.

Another example of cost savings from a unified numbering system can be obtained from centralized identification and control of specification control drawings. Several divisions within a company may obtain the benefit of a specification control drawing if a single system of identification is used. Thus duplication of specification control drawings can be reduced or eliminated by having a single control drawing identification and control system that applies to all the divisions. Of course, its effective operation requires a designer or controller who will check all new requests for a specification control drawing number to verify that a drawing is not already released for the item desired. The addition of such an individual to the staff may increase overhead costs, but he will save the company many times his wage if specification control drawings are prepared on a large scale.

The format and information contained in identifiers can vary within the company system to differentiate among the various kinds of documents. However, for each type of document, conformity of format is essential for efficient identification and use by personnel. This conformity should apply to all projects, large and small. Thus a specification control drawing identifier consisting of a prefix and six digits (ST 123XX6 or C 123XX6) should be the only authorized format unless the customer requires use of his own identification numbers.

12.1   KEY ASPECTS OF DRAWING AND PART NUMBERS

Drawings and parts are identified by alphanumeric coding systems. However, common features of most numbering systems are that drawing and part numbers are identical, or nearly so, and identifiers are assigned to all parts. When a part is purchased to an industry, federal, or military standard, the standard number is used for the part identification

---

[1] *Cataloging Handbook* H7, "Manufacturers Part and Drawing Numbering Systems for Use in the Federal Cataloging System."

number. Commercial or nonstandard parts are identified by their manufacturer's numbers. Two key rules to remember about part numbering are that (a) all parts that are physically and functionally interchangeable must be assigned identical numbers and (b) different part numbers cannot be assigned to new or modified parts if they are interchangeable with other parts already identified.

Following these rules helps to avoid excess inventories of spare parts and artificially created shortages. For example, if parts purchased from three vendors to a specification control drawing are identified by their vendor part numbers, these parts will be received and stocked as separate items because the material control clerk does not examine each part to determine whether it is the same as some other part so that it can be stocked in the same bin. As a result costs will increase for the separate inventory records and stock bins maintained for parts which are essentially identical. Besides the additional cost, artificial shortages will be created when the supply of one of the parts is exhausted. Consequently new purchase orders will be issued even though the second and third stock bins (identified with different numbers) may have enough parts for the project's needs.

The exact numbering system used depends on company requirements. However, there are general guidelines for developing a system. These guidelines are similar to those for good technical writing: accuracy, simplicity, conciseness, clarity, and completeness. In addition, a good numbering system is based on common-sense evaluation of your needs and the criteria listed below[2]:

Characteristic(s)	Value
Convenient, functional and adequate	Provides minimum effort to apply and satisfy all company needs.
Brief and simple	Reduces errors in copying and simplifies retention.
Versatile	Accommodates new types of items and documents without changing the basic system structure.
Distinctiveness	Permits easy recognition and retention.
Uniformity	All numbers have the same appearance simplifying identification and retention.
Integrated	All documents pertaining to the same item are related through a common number[3]
Nonsignificant	Numbers do not contain coded data in the form of order or types of letters given in the number.
Extra coding system	Distinguishes one design activity from another.

[2] Joseph Mazia and James V. Panek, *Numbering Systems*, 9th Annual Meeting of the Standards Engineering Society, Pittsburgh, 1960.

[3] An item identified as 123456 is integrated with its parts list by the number PL 123456, its wire list by WL 123456, and its test procedure by TP 123456.

Most of the criteria just presented are based on common sense and do not require further explanation. However, the use of nonsignificant numbering systems instead of significant ones needs some additional explanation. A significant numbering contains information in coded form as follows:

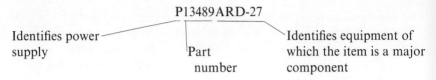

P13489ARD-27

Identifies power
supply

Part
number

Identifies equipment of
which the item is a major
component

The disadvantage of this kind of system is that it is cumbersome and inflexible. Also, it is not reliable when rapid technological changes occur and when the possible combinations required are so large that the system becomes saturated by using up the possible combinations available.

A part number is the principal configuration identifier for most documents and records dealing with parts. Whenever possible, these documents and records are sorted, filed, and classified by part number. While the eight characteristics listed before apply to both drawing and part numbers, the following additional features characterize a good part numbering system:

1. Complete and unique identification for each part or item.
2. Easy to handle in parts lists, drawings, records, correspondence, etc.
3. Readily adaptable to data processing equipment, thus avoiding the use of alphabetical digits with the part number and having the same number of digits in each part number.
4. Easy conversion from existing part number to new number; for example, old part number is changed from 12XX34-1 to 12XX34-2.
5. Classifies item, such as resistor, capacitor, screw, nut, etc. This can be done by prefixes, such as RC-1234X2, where RC identifies carbon resistors.
6. Avoids unnecessary part number changes because of changes in the part's application, cost, procurement source, etc. This is achieved by having a configuration review board that checks each proposed part number change to assure that it is valid.
7. Avoids assigning different part numbers to identical items and the same number to different items. Attainment of this feature is the same as for 6 above.

Note that features 3 and 5 conflict in that 3 calls for using pure numbers while 5 uses letters for indicating the type of items. The use of 3 or 5 must be made by comparing the benefits of each system in terms of what is best for the company and customer.

12.2 NUMBER COMPOSITION

As mentioned before, drawing and part number compositions vary widely depending on the company. However, the entire combination of numbers, letters, and dashes cannot exceed 15 characters, excluding revision letters, if government work is desired. Numbers larger than 15 characters present identification and perception problems for staff members and can glut automated data processing equipment. These alphanumeric identifiers have the following restrictions:

1. Letters I, O, Q, S, and X cannot be used because they can be confused with similar looking numbers or letters.
2. All letters shall be capitals.
3. Numbers shall be whole arabic numerals.
4. Blank spaces cannot be used.
5. Symbols, such as #, /, *, etc., cannot be used.

The drawing and part number can be represented by one number (mononumber system) or can consist of three parts: (a) basic drawing number; (b) revision letter (drawing only); and (c) suffix number. Note that the drawing number consists of a basic X-digit number and a revision letter. The part number consists of the same X-digit number minus the revision letter, plus a dash suffix number. In some systems, suffix numbers are issued to drawings also, although this practice should be discouraged because they may be confused with part numbers.

Drawing and Part Numbers

The first element of a complete drawing and part number is the company-assigned drawing number for the part or assembly. This number is a permanent element of the part number. It identifies the original usage or application and any future usage for all interchangeable or noninterchangeable versions of the part, component, subassembly, assembly, or equipment. Parts, components, subassemblies, and assemblies are referred to as items in the following text.

Once assigned, the drawing number is not changed in any way or reassigned to another drawing. Thus, the family relationship provided by the permanent drawing number for all noninterchangeable versions of an item, identified by a dash number suffix, is retained for all design applications within a single contract end item series or type, model, series.

When multidetail assembly drawings are used, the original dash numbers assigned to identify the several detail parts and subassemblies contained on the drawing are a part of the permanently retained part number. Although these dash numbers are used to identify the part, the drawing is only identified by the basic drawing identification number.

Revision Letters

Changes to the drawings are identified by revision letters issued in sequence whenever a change is made, regardless of its significance. Thus, the letter A is issued for the first change, the letter B for the second change, and so on. When all the letters of the alphabet are used up, the revision sequence can continue with a double letter combination: AA, AB, AC, AD, and so on. (Note that letters I, O, Q, and X are not used.)

Suffix Number

The second element of the part number consists of a dash and a dash number assigned by the design engineer when the original design application of any detail item shown on a drawing is made noninterchangeable with the previous design because of an engineering change. The dash number has the same characteristics as a drawing number and may consist of letters and numbers.

When detail assembly drawings with tabulated items are required (see Chapter 3), two sets of dash numbers are used for part identification. The first dash identifies the tabulated item and the last dash number always denotes a noninterchangeable condition.

If several items are shown on a single drawing, the identifying part number is the drawing number with a suffixed dash number. With parts identified this way, only the suffix number has to be listed in the parts list (also referred to as a list of materials) contained in the drawing that the part number originated from; the basic drawing number doesn't have to be repeated.

When detail parts that do not have individual drawings are shown on an assembly drawing, the parts are listed in the parts list of the assembly with assigned dash numbers. Assignment of a dash number is made so that it differentiates the dash number of an assembly from that of a detail part. A descending order of 3-digit numbers may be used to identify different levels of the product, although this amount of significant numbering is not recommended. For example:

Assembly	Assigned Suffixes
Top	400
Assembly below top	300 through 399
Subassembly	200 through 299
Component	100 through 199

Note that for small products that use only monodetail drawings, the basic drawing number equals the part number and dash numbers need not be used, except for tabulated parts drawings. When parts become noninterchangeable in this system, a new drawing number is assigned instead of a dash number.

Sample Drawing and Part Numbers

Three types of drawing and part numbers are shown below. These identifiers are purely numeric except for revision letters and do not contain significant data. Note that in each case the maximum number of characters, including dashes, does not exceed 15.

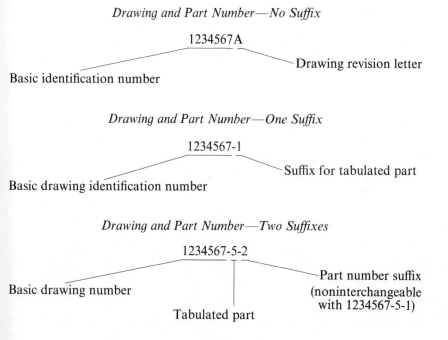

Drawing and Part Number—No Suffix

1234567A

Drawing revision letter

Basic identification number

Drawing and Part Number—One Suffix

1234567-1

Suffix for tabulated part

Basic drawing identification number

Drawing and Part Number—Two Suffixes

1234567-5-2

Part number suffix (noninterchangeable with 1234567-5-1)

Basic drawing number

Tabulated part

Family Designation Numbers

Family designation numbers (FDN) are used to identify parts and assemblies (items) below the end item level that are similar in configuration and are used for the same application; for example, an amplifier installed in each product of a series. The FDN is also used by some companies to identify all items and products required for a specific program. A sample identification number for this use is shown below:

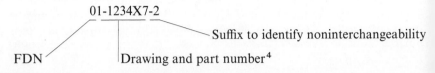

For this system, all item identification numbers for Project X, identified by 01, contain the prefix 01 (the prefix could be all letters or a combination of numbers and letters). The following discussion applies to the first use only, that is, to the use of FDN to relate all similar configuration items used in the CI series.

At the product level, the permanent CI number and its serial number provide individual identification of each product belonging to a specific design application, even though its part numbers change because of a noninterchangeable configuration. The FDN provides this same identification capability for items below the product level and relates these items to the design application or usage within a specific end item or product. Examples of these numbers follow:

Part Nomenclature	Identification Numbers

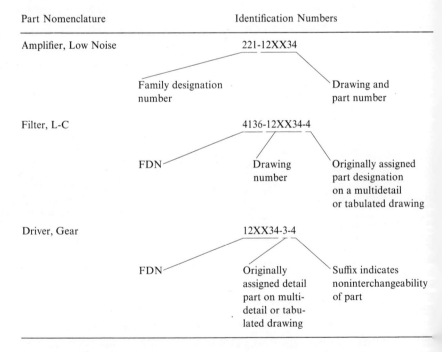

---

[4] Note that the drawing would be identified by 1234X7 and the part by 1234X7-2.

The application of the FDN system requires the use of dash numbers for identifying monodetail, multidetail, and detail assembly drawings and part numbers. (A detail assembly drawing shows items in their true locations in the assembly.) The multidetail system requires the use of two dash numbers, one to identify the detail part and the other to identify a noninterchangeable design. The principal reasons for using the FDN are to enable uniform identification of items to prevent the purchase or construction of more items than are needed, with resulting excess inventories, and to avoid item shortages. The functions of the FDN are listed as follows:

1. The FDN enables identification of a noninterchangeable item in the same family; thus, it provides the total number of items built for a product series and the number of items required for spares provisioning. Noninterchangeable items will each require spares, the quantity of which will depend on the number of noninterchangeable items used in the series.

2. The FDN provides a permanent base for serializing items when necessary to satisfy quality assurance and logistics requirements. The items are sequentially serialized with respect to this base even though their dash suffix number is changed to indicate noninterchangeability within the family. Once assigned, the item serial numbers are not changed even though the item is reworked to a new configuration. The FDN and sequential serialization simplify identification of items in the same family and facilitate systematic evaluation and rework of items requiring a changed configuration.

Family designation numbers do not have to be applied to all items. The FDN should be assigned when one of the following conditions applies:

1. Item is critical to CI operation.
2. Serialization is required.
3. Individual identity and accountability are required for end item application, manufacturing, and quality assurance records.
4. Item is of unique or original design.
5. Item is repairable.

Nonrepairable off-the-shelf vendor items usually do not require FDN's because next assembly number or CI number entries are not required in the drawing application blocks and effectivity data are not given for address to subordinate part changes. Items for which the company is the design manufacturer usually need an FDN. Exceptions are items that are permanently combined into an assembly and which cannot be removed, reworked, or replaced without destroying the assembly.

Family designation numbers rarely should be changed. The rule for changing an FDN is: If an item built to the superseded design is not capable of being reworked to the new design, the FDN should be changed. If a new drawing is required to show the new design, the FDN should probably be changed. However, if a new drawing is made to improve quality or obtain a better layout, the FDN or part number suffix should not be changed. Only a drawing revision letter is required for this change.

12.3 DRAWING AND PART NUMBER ASSIGNMENT

The basic drawing numbers are assigned to each project by data control. These numbers are usually assigned in blocks of 200 to 1000 numbers for each product. The design engineer assigns these numbers to drawings sequentially, beginning with the lowest number, usually assigned to the top assembly drawing which shows overall product configuration. He also assigns part numbers if they are different from the drawing numbers, based on the data supplied by the project engineers. As mentioned before, the design engineer assigns dash numbers to the basic drawing number when required to identify a noninterchangeable part. Once issued by data control, these numbers are not reused regardless of whether they have been assigned to a drawing or not.

The quantity of numbers in each block varies, depending on the project or company policy. As an example, assume that three blocks of drawing numbers are issued to Projects A, B, and C. The drawing blocks could appear as follows:

Project (in order of request)	Drawing Number Blocks	Total Quantity of Drawing Numbers for Each Project
A	280000 through 280999	1000
B	281000 through 281999	1000
C	282000 through 282999	1000

The data control clerk is responsible for recording the number blocks issued to each project, the project title, contract number, the person receiving the numbers, and the date of issue. Once assigned, the drawing block cannot be reissued under any conditions, including cancellation of the project. The risk that the same number may appear on two drawings is too high. Of course, when all the numbers have been assigned after several years, the sequence can begin again with the first assignment made.

Once a block of numbers is issued to the designer, individual numbers from that block are issued by him as preparation of each new drawing begins. The designer keeps a drawing log which identifies each drawing number by title, originator, and date of issue as shown in Figure 17.1. Upon completion of the project, this log book is turned into data control for safekeeping and future reference.

12.4 CHANGING PART NUMBERS

After a part number is assigned, only the suffix of a production part number may be changed. Changes to the suffix are required when the new part is not physically and functionally interchangeable with previously built parts. Exceptions to this rule and the details for reidentifying part numbers are described in the following sections.

Change Letter Reidentification

Engineering changes can be incorporated and identified by drawing change letter control, without changing the part number, up to the cut-off date for incorporation of changes on the first production article to be accepted by the customer. A revision letter change system such as this is acceptable if all such changes are made effective on product serial number 01 and on. After the first production product is accepted, drawing change letter control continues for minor changes but part numbers are changed when interchangeability is affected as described below.

Part Number Reidentification

The dash number or part number suffix is changed by the designer, with the configuration manager's approval, whenever one or more of the following noninterchangeable conditions occurs:

1. Items are altered so that old and new parts are not directly and completely interchangeable with respect to installation or performance.

2. Old items are limited to use in specific serial numbered equipment but new items can be used in any equipment.

3. Performance or durability is affected to such an extent that old items must be scrapped for safety or because of poor or faulty operation or low reliability.

4. The company alters or selects the item that has been identified and documented by another company. In this case, a new drawing and part number is assigned and the following information is added to the drawing:

   a. The original drawing and part number.
   b. The manufacturer and his code identification number.
   c. Manufacturer's address.

5. A material, process, or protective treatment is altered so that any of the conditions described above exist.

6. An item is reworked, using a modification kit, into a later dash number version of the item and is completely interchangeable with all items identified by the later dash number (the part is identified with the dash number of the later revision).

7. An item is established by the company as a standard and identified by a standard specification identification number when the following conditions apply:

a. The item has a multiple usage and is expected to be used in more than one end item.

b. The item is not repairable and spares will not be kept by the customer below the level identified by the standard specification identification number.

c. The item is completely defined in a specification, source control drawing, or specification control drawing, including performance, durability, reliability, form, fit, quality, and inspection requirements.

d. More than one source is approved and qualified to supply the item.

8. When interchangeable repairable assemblies contain noninterchangeable items, both the noninterchangeable items and their parent assemblies change identification.

Changes to Higher Level Assembly Part Numbers

When a repairable assembly contains a noninterchangeable item, both the noninterchangeable item and the parent assembly are reidentified with new part numbers. In addition, all progressively higher assemblies up to and including the assembly where interchangeability is reestablished are reidentified. Part numbers are not changed above this level of assembly for any reason. (When part numbers are unnecessarily changed above this level, the process is called part number tumbling.) Reidentification is usually accomplished by the use of dash numbers.

An example of reidentification of higher level assemblies resulting from two lower level noninterchangeable alterations to a piece part is depicted below.

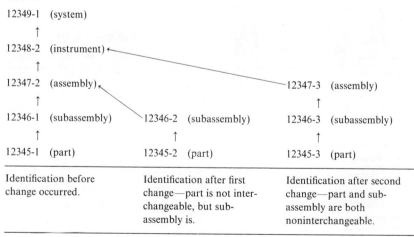

| Identification before change occurred. | Identification after first change—part is not interchangeable, but subassembly is. | Identification after second change—part and subassembly are both noninterchangeable. |

Assuming arbitrary serial number effectivities, the following conditions exist as a result of the above changes: (a) subassembly 1 is used in assembly 2, serial numbers 1–10; (b) subassembly 2 is used in assembly 2, serial numbers 11–25; (c) assembly 2 is used in instrument 2, serial numbers 1–25; and (d) assembly 3 is used in instrument 2, serial numbers 26 and subsequent.

Changes Not Requiring Part Number Reidentification

In many cases, part numbers do not have to be reidentified because of a change in the part's design. Remember that the principal reason for changing a part number is to indicate noninterchangeability of form, fit, or function. When a minor change is made to the drawing describing the part, the drawing is reidentified with a revision letter, but the part number remains the same.

The drawing and part numbers do not have to be changed when the following four conditions exist:

1. A new usage is found for an existing part.
2. None of the conditions listed under *Part Number Reidentification* exist.
3. A commercial or government furnished item is used without alteration or selection.
4. A commercial item is established as a company standard without alteration or selection. However, the specification control drawing number is used to identify the part in drawings and parts lists.

12.5 SPECIAL PART AND DRAWING NUMBERING SITUATIONS

Some special situations related to part and drawing numbers are described below. These include identification of altered, selected, matched, and symmetrically opposite parts by drawing numbers and drawing notations. The policies selected by the company should conform to these requirements, and procedures for implementing these policies should be released to all design and drafting personnel.

Altered Part Identification

An altered part is a vendor item that has been changed in some way. For example, hole sizes of a relay mounting plate may be increased to allow the use of larger bolts. When a vendor part is altered, a drawing is prepared showing the part and describing the alteration. The altered part is then identified in parts lists and other documents by the company's drawing number. In addition, the vendor, his code identification number, and the original part number are recorded on the field of the drawing, usually to the left of the title block. Further identification is provided by printing "Altered Item Drawing" above or near the title block.

Selected Part Identification

Occasionally, parts must be selected for a special application. A critical circuit may require a 2.5 ohm resistor which has a resistance value that is within 2 percent of its nominal value. If the manufacturer only guarantees an accuracy of 5 percent, the company must order a quantity of these resistors, measure each one, and select those with resistances between 2.45 and 2.55 ohms. Of course, these resistors must be identified, packaged, and stored for later use in the circuit during manufacturing.

A special drawing is usually prepared for identification of the selected item. As with the altered part, the drawing represents the new part number and records the original manufacturer and part number. Instructions for selecting the part are added to the drawing next to the vendor identification. "Selected Item Drawing" is printed above the title block.

Matched Set Identification

A matched set or pair of items is sometimes required for an equipment. These items, such as four interrelated gears that must mate with minimum binding and backlash, are depicted on one drawing. As with the altered and selected items, the drawing number becomes the new part number and is used to build or reprocure additional gear sets. Since mated parts are always replaced as a set, a single part number must be used for their identification.

A box is placed on the drawing to identify each item of the set and contains a note that the parts comprise a matched set. The data are recorded as follows:

These parts constitute one matched gear set

Dash Number	Part Number	Quantity/Set
-1	123X45	1
-2	423X56	1
-3	344X11	2

If vendor items are used, the vendor and his code number must also be identified. Assembly drawings, parts lists, and other documents should identify the set only by the number of the new drawing, not by the individual part numbers or dash numbers. The entry "Matched Set" should appear above or to the left of the title block.

Symmetrically Opposite Part Identification

Symmetrically opposite parts are items that are mirror images of each other. When symmetrically opposite parts or assemblies are required for the equipment, one of the parts is shown on the drawing and is identified by an odd dash number. The opposite configuration of the part, which is not shown on the drawing, is identified by the next even dash number A table is added to the drawing listing the dash numbers and identifying the two configurations. For example, the table would include: 456XX5-1 shown and 456XX5-2 opposite. The basic drawing number is 456XX5. Odd dash numbers for shown parts is the preferred identification method. The space above or to the left of the title block may be identified with "Symmetrically Opposite Parts."

Specification and Source Control Drawing Identification

Parts that are used without alteration or selection are identified by the vendor part number. However, specification or source control numbers may be assigned for material control and to simplify references to the parts by using a single identification number. Specification control drawing numbers are really control and accounting identifiers and not part numbers.[5] Therefore, when the specification control drawing number is given, the parts list should contain the following note in an adjacent column: "Vendor Item—See Specification Control Drawing." Note, however, that source control drawing numbers can be used as part numbers. The following note is also added to source control drawings:

"Only the item described on this drawing when procured from the vendor(s) listed hereon is approved by company AS, Pasadena, California, for use in the application specified hereon. A substitute item shall not be used without prior testing and approval by the AS company or by NASA."

Substitute Part Identification

Substitute parts are two or more parts that possess similar functional and physical characteristics allowing them to be exchanged under certain conditions or in special applications without alteration of the parts themselves or adjoining items. A part is identified as a substitute part even though only one way interchangeability exists; that is, where part B can be interchanged in all applications for part A, but part A cannot be used in all applications requiring part B. Substitute parts are identified by different

[5] NASA allows the use of specification identification numbers for a nonstandard part number until the vendor part is selected. Then the vendor part number must be recorded in the "Item Identification" column of the parts list.

part numbers. When a substitute part is used, this should be noted in the "Remarks" column of the parts list and other applicable documents. If some characteristics are different from the original part or special conditions apply, the different characteristics and special conditions should be described.

Replacement Part Identification

A replacement part is functionally interchangeable with another but differs physically from the original part because installation of the replacement part requires special operations in addition to the normal methods of attachment. Special operations include drilling, reaming, cutting, filing, and shimming. Replacement parts are identified with numbers different from those of the parts they replace. Parts lists and other documents should contain a note stating that a replacement part was used and how it differs from the original part.

Inseparable Assembly Part Identification

An inseparable assembly consists of two or more items that are separately built and permanently joined together by riveting, welding, or the like, to form a single assembly. This assembly is shown on a separate drawing and is assigned its own part number. The items which make up the inseparable assembly can be shown on separate drawings and identified separately from the assembly.

Government and Industry Part Identification

Government and industry parts retain their original numbers and are not identified by the company part number unless modified so as to require a new identification number. The criteria for issuing new identification numbers are the following:

1. The previous characteristics of the part no longer enable it to be qualified for use on the equipment.
2. Previous selection or inspection standards have been changed to such an extent that new quality inspection standards are required.
3. Specified characteristics have been changed so much that reworking or reidentification of production items is required.
4. A new characteristic has been added to the part.

If an item is identified by a vendor or other source with an identification number that exceeds 15 characters, a control number may be assigned to meet customer requirements limiting the length of the number.

12.6   INTERCHANGEABILITY IDENTIFICATION

The determination of whether a proposed change will leave an item interchangeable with its original configuration requires conscientious and objective treatment. Correct designation is of major importance to good configuration management because it affects part number control, stocking of parts, inspection procedures, repairs, and recovery of fee. In a marginal case, where reidentification is not clearly required, the forces at work may affect the personnel involved in the decision as follows:

If it is up to the design engineer, he will probably want to minimize the number of drawings by keeping the part number unchanged so that the part is interchangeable. The material control and manufacturing engineers will agree with the design engineer because they want the parts bins and planning paper as uncomplicated as possible. On the other hand, the quality assurance engineer will insist on calling a "spade a spade," as will the product support engineer who no doubt has had sad experience with replacements which were identified with the same number but were in fact noninterchangeable. Finally, the contracts administrator will be torn between a desire to collect fees for added work and a tendency to avoid borderline in-scope changes.

To help resolve the difficulties that arise from the differing opinions described above, the following concepts and rules should be reviewed before making the final decision on whether reidentification of a part number is required. (Note that most of these items were discussed in previous sections but are given here because of their importance in determining the interchangeability of parts.) Items are interchangeable under the following conditions:

1. When they can be freely exchanged with each other regardless of application. But an item is considered noninterchangeable when it is used in more than one location on one CI or on more than one CI and a change is made that affects the physical or functional application of the item in any of the locations.

2. When they can be exchanged one for another without alteration of physical installation provisions and without alteration of connections (electronic, electrical, hydraulic, pneumatic, or mechanical).

   a. An item is considered noninterchangeable if a change is made that affects the bracketry on which the item is mounted, increases the space envelope allocated, or changes the permanently installed fasteners or the fastener hole pattern.

   b. An item is considered noninterchangeable if a change is made that affects the design or installation of push-pull rods, jackscrews, limit switches, etc.

c. An item is considered noninterchangeable if a change is made that requires different connections, such as tube and pipe fittings or electronic and electrical connectors, or that adds wiring to open pins in the existing connector.

3. When they can be exchanged one for another without loss of any function or change in performance of the item or its system.

a. An item is considered noninterchangeable if a change is made that adds a function which the earlier manufactured items will not perform.

b. An item is considered noninterchangeable if a change is made that results in a change to the acceptance test specification and the earlier manufactured items will not meet the new requirements.

4. When they can be exchanged one for another without changing preset adjustments or adjustment schedules.

a. An item is considered noninterchangeable if a change is made that requires changing the preset adjustment of earlier manufactured like items or any other related items.

b. An item is considered noninterchangeable if a change is made that requires changing adjustment schedules, such as fuel transfer for center of gravity adjustment.

5. When they can be exchanged one for the other without affecting the signal or voltage output of the module, assembly, CI, or system.

a. An item is considered noninterchangeable if a change is made that causes the magnitude of the output signal to change and produces an out-of-tolerance condition or detrimental effects to any associated circuits.

b. An item is considered noninterchangeable if a change is made that causes the magnitude of the output signal to fall outside the tolerances of the earlier manufactured item or causes any associated circuits to operate outside their established tolerances.

6. When they can be exchanged one for another without modification of test sets or test equipment. But an item is considered noninterchangeable if a change is made that requires any change in a delivered test set or test equipment.

7. When they can be exchanged one for another without creating the need for more than two computer tape configurations for a given test set or test equipment.[6]

[6] The preceding section was adapted from notes supplied by Walker Bennett, Nov. 1969.

Chapter 13

CHANGE PROPOSAL IDENTIFICATION NUMBERS

Change proposal identification numbers are the key configuration identifiers for identifying and controlling engineering changes and resulting administrative actions requiring customer approval. These numbers are the only ones authorized for contract or end item identification of the change status of an equipment and for reporting the state of the actual or in-process configuration to the customer.

The salient feature of change identification numbers is that they uniquely and completely relate an engineering change to individual serial numbered equipments being produced by the company. Therefore, to avoid ambiguity, only one change is allowed to be identified by each change identification number. By using a single number for a single change, the problem of having one change approved and another rejected is avoided, as would occur if more than one change were identified by the same number.

The quantity of change identification numbers issued during the project depends on the number of changes to be proposed to the customer, which in turn depends on a variety of conditions, such as the success of the design and development effort, changing customer requirements, schedule conflicts, or new technological developments outdating the design. The format and length of the number are usually determined by the customer because he has to process the engineering change proposal with his internal change identification and control system. Failure to comply with his requirements may result in his refusal to review the ECP, bad feelings, or delays in processing and approval. In addition, he may assign his own number to the ECP, which will be recorded on a contract authorization to proceed with the change. If the authorization document does not include a cross reference to the company's change identification number, the company may relate the authorization to the wrong ECP. The consequences of this type of error can only result in rework, schedule slippage, and incorrect configuration status records.

Vendors or small subcontractors may not be required to submit change proposals for engineering changes. Instead, engineering orders may be prepared and submitted to the company with a transmittal letter indicating cost and schedule changes that will result from the engineering change described on the EO. In this case, the EO number is the change identification number.

13.1   ECP CHANGE IDENTIFICATION NUMBERS

The change proposal identification number (see Figure 5.6) is a packaging number assigned by the configuration manager to engineering data describing an engineering change. The change identification number applies primarily to Class I engineering changes but may also be used for Class II changes when the customer requires that he approve the change before it is made. The identification number is used to control, sequence, and account for production and retrofit actions resulting from the change. (Refer to Chapter 2 for a detailed description on the classification of changes.) Engineering change proposals are identified with numbers consisting of three basic elements: prefix, sequence number, and correction code.

1. *Prefix* relates the change to the equipment, company, and system affected by the change.

2. *ECP sequence number* identifies each Class I or II engineering change in the sequence of preparation. The ECP sequence number identifies the basic increments used for configuration control and accounting by the customer of the approved equipment configuration. In addition, a code letter is used to identify the nature of the ECP: preliminary, formal, etc.

3. *Suffix correction code* distinguishes among ECP documents containing major revisions, minor editorial corrections, and the original submittal.

The format of the change identification number for an ECP is shown below:

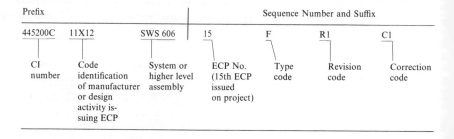

This example identifies the first revision to the fifteenth change to equipment 445200C processed by the company with identification code 11X12 for the SWS 606 system. In addition, a minor correction was made to revision 1. The equipment model and system designations are defined at the beginning of the project and are therefore identical for all ECP's prepared during the project.

The configuration manager assigns ECP numbers consecutively from a separate series for each project. The first ECP number is identified with 1, the second with 2, the third with 3, and so on. Once assigned, the number is retained for all subsequent submissions of the ECP and for all ECP types, revisions, and correction codes that may be added to the number.

Engineering changes that affect more than one type of equipment are identified by a basic ECP number with a separate dash number assigned for each equipment affected. For example, if a radio (one equipment) requires an engineering change that also affects a related transmitter (a second equipment) and a test set for checking both units (a third equipment), the three equipments would be interrelated as follows:

Equipment	ECP Number
Radio (receiver)	ECP 1200
Transmitter	ECP 1200-1
Test set	ECP 1200-2

Each dash-numbered ECP must be complete for the equipment change being proposed. If an equipment being built by another company is affected, the project manager must coordinate the change with that company's project manager so that an ECP can be prepared by him. Of course, the ECP number assignment and preparation will be performed by the other company. The results of the coordination effort and the other company's ECP number, if available, should be submitted with the ECP package to the customer.

The configuration manager may assign one of the two types of code to each ECP: preliminary or formal. The code may be changed to reflect a new ECP condition whenever it is necessary to resubmit the proposal. Of course, changing the ECP to a formal submittal requires a revision change. A more complicated system of coding is sometimes used. The code designations for this system are given in Table VII.

Revisions to ECP's are made when a major change is made to the ECP. The configuration manager identifies each formal revision to the ECP with an "R1" for the first revision, "R2" for the second revision, "R3" for the third revision, "R4" for the fourth revision, "R5" for the fifth, and so on. He also identifies minor editorial changes to the ECP with an R1-C1

TABLE VII
ECP Type Descriptions

Code	ECP Type	Description
P	Preliminary	Initial data submitted to customer on a change being considered by the company. Continued work on ECP is halted until customer approval to go ahead with analysis.
F	Formal	Follows preliminary ECP with intent to achieve contractual authorization.
S	Secondary	Applies to an ECP prepared by a company or subcontractor having a CI contract with the customer but not having prime design responsibility for the equipment.
C	Compatibility[a]	Applies to system/equipment changes required to make the system/equipment work during functional checks, installation, and checkout.
E	Expedite[a]	Applies to cases requiring emergency (immediate action) or urgent (processed within 48 hours) action.
R	System requirements	Applies to a change to an approved performance and design requirements specification for a system.
D	Design requirements	Applies to a change to Part I of the CI specification after the specification has been approved.
X	Integrating	Determines impact of a change on all elements of a system, including effects on system requirements, interfaces, and CI's other than those of the ECP's originator.

NOTE: The above codes are the most familiar ones used in industry. However, MIL-STD-480, which supersedes previous ECP preparation instructions, uses different codes.

[a] Routine ECP is any ECP that does not have an urgent, emergency, or compatibility priority. Usually processed in six weeks.

for the first correction to the first revision; an R2-C1 is the first correction to the second revision. Unlike the revision letter, the correction code is not recorded in the status accounting records because it does not affect the equipment configuration.

The ECP sequence number (15-1 P R2) is the number that is used in configuration indexes and accounting reports to identify the ECP. Consequently, the customer may restrict the total number of characters for numerals, dashes, and letters to a quantity that can be easily handled by his data handling system. Some government agencies restrict the sequence number to 11 characters.

13.2   CLASS II CHANGE IDENTIFICATION NUMBERS

Class II change identification numbers not requiring ECP's are constructed in accordance with the company's internal system and procedures if these procedures are compatible with the customer's requirements. These numbers usually consist of a prefix and a 4-digit sequence number as follows: EO 1002. The next EO written on the project would be assigned EO 1003. No revision letters are assigned to EO numbers. If an EO is to be changed, a new sequence number is issued for cancelling the old EO and describing the change. As with ECP change identification numbers, once a number is assigned, it cannot be reissued even if the EO is cancelled or rejected.

13.3   DEVIATION AND WAIVER CHANGE IDENTIFICATION NUMBERS

In addition to ECP and EO numbers, change identification numbers are issued to deviation and waiver request documents. These documents describe temporary changes from approved document (drawings, specifications, and so on) requirements and are used to differentiate between a proposed change that will improve the equipment (ECP) and one that usually reduces the quality of the equipment's configuration (waiver or deviation).

Like ECP's, deviations and waivers are each identified with a separate series of sequential numbers. The change identification numbers are similar to an ECP, as shown in the following example:

Deviation Sequence	Model/Type	Manufacturing Code	System Designator	Deviation Number
First	445200C	11XX2	SWS 606	1
Second	445200C	11XX2	SWS 606	2
Third	445200C	11XX2	SWS 606	3

Waiver Sequence	Model/Type	Manufacturing Code	System Designator	Waiver Number
First	445200C	11XX2	SWS 606	1
Second	445200C	11XX2	SWS 606	2
Third	445200C	11XX2	SWS 606	3

Note that the code, revision, and correction identifiers are not used in deviation or waiver change numbers. Since deviations and waivers are usually one page documents that are explicit and complete change requests desired by the company, they are not submitted for preliminary approval by the customer. Revisions are also rarely made to these documents and revision codes are therefore not normally included. If a revision is found necessary, a new document may be issued with a new number or a change letter may be added to the original number:

445200C 11XX2 SWS 606 1A

revision letter

Also, the note "Supersedes Waiver No. 1, July 10, 1971," should be added below the waiver number or in the upper right-hand corner of the page if there is no room near the identifying number.

Chapter 14

DESIGN REVIEWS

Design reviews are a series of formal meetings held internally by the contractor during the design process. The purpose of these meetings is to critically examine the product design, configuration, design documentation, test program planning, and test data, and in specific cases the customer participates. A typical sequence of design reviews is shown in Table VIII. Refer to Figure 14.1 for time relationships.

Overall objectives of these meetings are to ensure that all performance and reliability requirements will be met and that no design weakness exists that will compromise the performance, reliability, or quality of the equipment. Specific objectives of design reviews are the following:

1. To achieve the best design approach possible by reviewing critically all major electrical and mechanical aspects of the equipment.

2. To identify and confirm the final product design and quality base for building flight production equipment.

3. To avoid errors in design and fabrication found in other similar equipments already built.

A principal function of design reviews is to provide customer and company management with data for determining the design status, identifying problem areas, approving production of hardware, conducting trade-off studies, requesting changes to specifications, and establishing test programs involving the different types of equipments that are combined to form a system.

Design reviews can range from very simple, straightforward affairs to complex and expensive operations involving dozens of people. The results of these reviews can be numerous tasks to be completed, extensive data preparation, and reports on the status of action items issued to correct hardware or data deficiencies. More complex design reviews may require advance preparation beginning at least one month before the meeting and continuing for a month or two after the review is completed. Of course,

TABLE VIII
Internal Project and Contract Established Design Reviews

Internal	Contractual
a. *Conceptual design review* Verifies the validity of the design and interface concepts at the system and subsystem levels.	a. Not applicable.
b. *Preliminary design review* Verifies the validity of CI preliminary design and engineering model programs.	b. *Preliminary design review (PDR)/ System functional audit (SFA)* Confirms the contractor's design approaches as optimum with respect to the product's system specification and supporting CI specifications.
c. *Development design review* Verifies the validity of internal features of the qualification configuration design and qualification test program.	c. Not applicable.
d. *Preproduction design review* Verifies qualification test results and the completness and accuracy of engineering data to be released to manufacturing.	d. *Critical design review (CDR)/Functional configuration audit (FCA)* Compares the contractor's completed production release data and qualification test data with respect to the individual CI specifications.
e. *Shipping configuration review* Final review of the physical configuration of the "shipping" CI and supporting manufacturing/test data.	e. *First article configuration inspection (FACI)/Physical configuration audit (PCA)* The formal examination of the as-built configuration of a CI against its technical, manufacturing, and test data.
f. Not applicable.	f. *Final configuration review (FCR)/ Flight readiness review (FRR)* Usually held at operational site to verify installation and checkout status of the product and to confirm its release into operational status.

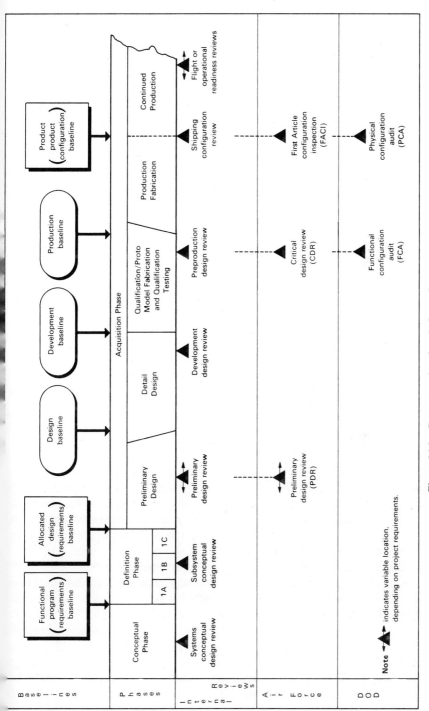

Figure 14.1 Contract and project design review sequence.

this effort is concurrent with the routine tasks of design, development, testing, and fabrication. Design reviews are enormously important to the successful design, development, and production of new equipment because in preparing for the review both the company's and customer's engineers must challenge and justify all key aspects of the design. Thereby they may detect any serious potential defects before production or field operation of the equipment.

Some of the areas examined during these reviews are the following:

1. Correct parts applications, including adequate derating and safety factor margins.

2. Alternative designs that may be simpler, cheaper, more reliable, or better performing.

3. Performance characteristics.

4. Reliability analysis.

5. Ease of fabrication, maintenance, and servicing.

6. Ability to survive exposure to the anticipated environment: vibration, launch, shock, salt spray, high humidity, high vacuum, fungus, micrometeoroid impacts, and so on.

7. Need for redundant circuits or components for critical areas.

8. Interface compatibility with other equipments in the system.

9. Identification and evaluation of critical failure modes.

10. Electromagnetic or mechanical interference with other equipments in the system.

Remember that preparation for the design reviews occurs during the entire project and not for just the four weeks preceding each meeting. The purpose of the four weeks before the design review is to schedule, organize and collect data to substantiate what has already been done during the project effort. If the requirements have not been built into the project all along, then they cannot be put there during preparation for the review. If a company is finding that many tasks have to be completed on an emergency basis before the meeting, then management should review project operations and prepare new guidelines and procedures for eliminating these deficiencies.

Another important point to keep in mind is that satisfactory completion of each review meeting results in customer or project authorization to proceed to the next phase of the project. Failure to satisfy the customer may result in schedule delays until the deficiencies at each stage are corrected. For example, in some projects a written directive from the customer is required before the company can begin prototype fabrication. Even though the parts, facilities, and personnel are ready to go, these cannot legally be put into use until authorization is received by the contracts administrator.

Any delay at this point results in wasted or unproductive labor and unnecessary standby expenses that must inevitably cause potential cost overruns and delivery schedule problems.

An extremely important aspect of design reviews is the corrective action items that result. These items are usually short tasks that must be completed by a project member to correct a design, hardware, or data deficiency. The date for completing the action item is always recorded, and an individual is given responsibility for overseeing all the action items and personally reviewing their status. When an action item is not being completed as planned, he follows it up by determining the problems involved and by taking steps, such as assigning additional personnel, to expedite its completion. The following are examples of action items:

1. To redesign an amplifier to provide higher gain because its performance is marginal with the current value.
2. To revise a procedure to incorporate additional steps for a test.
3. To find a more reliable transistor for a circuit.
4. To add external test points to the equipment.
5. To add a redundant switching circuit to a key component because of its critical nature to the successful operation of the equipment.

The minutes of the meeting must be carefully and completely recorded to ensure that an accurate record is available describing the decisions made, topics discussed, problem areas, action items and persons assigned responsibility for them, areas of agreement and disagreement, and administrative data, such as date and hours of meeting, location, attendees, and special breaks for small conferences or inspection of hardware. All participants should receive a copy of the minutes within a few days after the meeting. Another reason for the accurate and complete recording of design review results is that they are used to prepare the design review report.

Design review reports are prepared by the project manager or chairman after each design review. They are usually completed within two weeks after the design review and distributed to all design review participants. Reports are usually signed by the chairman and persons responsible for review actions; they include the following information:

1. A list of people who attended the design review.
2. The decisions or results of the review.
3. Corrective action items and responsible persons.
4. Schedule for completing the action items.

The following sections describe the personnel, design review board members, review data package, general conduct of design reviews, and each of the three design review types: preliminary, critical, and first article configuration inspection.

14.1   KEY PEOPLE

The project manager and design review coordinator are the key design review people. The project manager, or his designee, is responsible for the overall planning and scheduling of the reviews. He must assure that adequate preparation time, funding, and manpower are available for each review and that the requirements and objectives of each review are clearly defined. A design review coordinator is appointed by the project manager to organize, coordinate, and help administer the design review. His responsibilities include (a) promoting, coordinating, and supporting design review activities, (b) providing detailed instructions, (c) preparing schedules, (d) supporting all participants of the review team in technical and administrative areas, (e) coordinating with the customer and groups outside the immediate project team, such as publications, inspection, and manufacturing, and (f) recording and publishing the results of the design reviews.

14.2   DESIGN REVIEW BOARD

A design review board is set up to critique or review the design and equipment. This board is an advisory group that makes recommendations for improving the design or equipment. The review board consists of a chairman, cognizant engineer, and several members in specialized technical areas. The function of each group or member is as follows:

Chairman.   The chairman is the project manager or his project office representative. The chairman is responsible for the following tasks[1]:

1. Conducting overall design review and schedule.
2. Distributing review meeting notices.*
3. Preparing agenda defining the topics to be covered.*
4. Selecting experienced board members and assigning design review responsibilities.
5. Conducting the meeting.
6. Ensuring the recording and distribution of meeting minutes and maintaining files of design inventories and review board records. (Design inventories are written analyses of new or changed designs current at the time of the review.)*

[1] On many projects, the chairman is the customer's representative. Where a review board technical secretary is appointed, the items designated by an asterisk fall under his responsibilit

7. Taking action to implement board recommendations and to document the end results of these actions (generally referred to as the disposition of the action items).*

Cognizant Engineer. The cognizant engineer is selected by the project manager because of his thorough and detailed understanding of the technical aspects of the equipment and its design. He is responsible for assembling the design inventory, providing copies of technical data to the chairman, presenting the design approach to the review board, and answering technical questions. The cognizant engineer also plays a major role in completing technical action items that result from the review.

Board Members. Board members are specialists in design, testing, producibility, quality, reliability, logistics, and so forth, who have not themselves contributed directly to the documented design decisions. Each board member reviews and evaluates the area of design that he has been assigned to by the chairman; the chairman is responsible for providing review board members with adequate technical and design data for facilitating study of the design—electronic, mechanical, and thermal—before the review is conducted.

14.3   DESIGN REVIEW DATA PACKAGE

A design review data package is a collection of documents describing the design, analysis, and technical features of the equipment. The information and material required for the data package are prepared in sufficient detail to justify and describe the design for the review board. Data that should be provided are listed below. (When key areas have not been studied, they should be identified as incomplete.)

1. Description of the equipment.
2. Equipment layout.
3. Block diagram of the equipment.
4. Design requirements list (performance, environmental, and reliability parameters).
5. Specification list (identifies all specifications and standards used on the project).
6. Key drawings and schematics.
7. Wiring diagram.
8. Parts lists.
9. Drawing list.
10. Test requirements.
11. Lists of components, materials, and processes not covered by existing specifications or standards.

12. Functional and environmental test results.
13. Performance analyses.
14. Reliability prediction.
15. Cost analysis.
16. Alternative designs considered.
17. Weight analysis.
18. Transportation requirements and special handling techniques.
19. Specification control drawings for nonstandard parts.
20. Tolerance requirements studies.
21. Minimum safety margins based on stress analyses.
22. Design calculations.
23. Maintainability analysis and support evaluation, including ground handling and field testing requirements and safety precautions.
24. Known and potential problems.
25. Specification incompatibilities.
26. Power consumption analysis and profile.
27. Mission constraints and requirements.
28. Test plan.
29. Handling restrictions.
30. Ground support equipment description.
31. Range safety report.
32. Identification of interfaces.
33. Packaging considerations and techniques.
34. Center of gravity calculation.
35. Static load analysis.

14.4 GENERAL CONDUCT OF DESIGN REVIEWS

The general flow of a design review meeting is as follows: The review is scheduled as a half day session and usually begins in the morning to allow continuation of the meeting in the afternoon if necessary. Half day sessions are suggested because longer reviews are likely to lose their effectiveness. The second half of the day can be used to hold separate and small gatherings on an informal basis to resolve problems raised in the morning.

The meeting is usually started off by the department or division manager or chairman who presents an introduction to the review meeting, introduces the review board and project staff members to each other, and defines the objectives and scope of the review. Review board design considerations are given in Table IX. The responsible engineer should then lead the meeting by following the agenda and rapidly describing the equipment or design approach, using a block diagram to identify the key functional blocks of the

TABLE IX
Review Board Design Considerations[a]

1. Functional performance and stability
2. Component interchangeability.
3. Functional life
4. Reliability apportionment, prediction, and assessment
5. Cost
6. Value[b] engineering
7. Manufacturing, special tooling, special processes
8. Equipment outline and shape
9. Mounting and installation characteristics
10. Radio noise and interference susceptibility
11. Materials, parts, and processes used
12. Weight
13. Stresses applied to components (mechanical and electrical)
14. Relationship and compatibility with other equipments
15. Interfaces
16. Quality provisions
17. Test provisions and data
18. Maintainability and serviceability
19. Support requirements (manuals, spares, and so on)
20. Environment
21. Safety
22. Tolerance studies (e.g., what is allowable tolerance buildup caused by aging, high temperatures, and electrical stresses)
23. Failure mode, effect and criticality analyses
24. Alternative approaches
25. Associated documentation
26. Trade-off studies
27. Redundancy
28. Storage and transportation
29. Operator training
30. Test equipment
31. Field support
32. Identification of unreliable parts and time-limited parts

[a] The order of importance depends on the customer and contract requirements.
[b] Methods for reducing costs by simplifying the design or eliminating unnecessary requirements.

equipment. The description is followed by a question and answer period and action items are recorded.

Responsibility for following up action items resulting from the discussions is assigned to people at the meeting by the chairman, and their assignments and completion dates are recorded in the minutes of the meeting. During the afternoon the group may break up into small groups to continue the review of specific technical areas and to search for possible solutions to problem areas.

A copy of the minutes of the meeting should be given to the customer after the meeting and the review report is normally sent to the customer within two weeks (when required). The review is not complete until all action items are completed and the company is officially notified by letter or teletypewriter exchange that the review has been satisfactorily closed out.

Note that two or three mornings may be scheduled for the review to examine all the important areas and to resolve special problems. The subjects to be considered during the contractual design reviews are listed in Table IX.

14.5 PRELIMINARY DESIGN REVIEW (PDR)/SYSTEM FUNCTIONAL AUDIT (OPTIONAL)[2]

The PDR/SFA review is a formal customer review of the basic equipment design to assure the acceptability of the engineering approach. The engineering approach is evaluated by reviewing specifications, drawings, analyses, test data, and interface requirements. Available hardware such as mockups (simulations of flight CI shape, size, and weight) and breadboard models are also examined. A satisfactory design results in acceptance of the design and performance specification for the equipment. (This is usually referred to as Part I of a 2-part detail CI specification. Part II is approved at the completion of the last of the three major design reviews, first article configuration inspection.) A key aspect of the PDR is the identification of documentation that defines the physical and functional interfaces of the equipment with the system or other equipments.

14.6 CRITICAL DESIGN REVIEW (CDR)/FUNCTIONAL CONFIGURATION AUDIT (FCA)[3]

The critical design review (CDR) is a formal customer review of the equipment's detail design and is conducted to give the company formal approval to proceed with the manufacture of the equipment. The CDR is held when the detail design is about 90 percent complete with respect to formal engineering release for manufacturing and all interface control drawings have been agreed to by all affected contractors. Interface is defined as the boundary conditions and requirements for enabling various equipments to fit together without interfering with each other or with the overall system of which they are a part.

[2] PDR is AFSCM 375-1 terminology.

[3] AFSCM 375-1 terminology. FCA is the DOD standard terminology recently established by DOD Directive 5010.19.

The objective of the CDR is to verify that the detail design and performance characteristics of the equipment correspond with the data presented in the drawings, specifications, test reports, and mockups or engineering models. (An engineering model refers to an equipment that is built under laboratory conditions without manufacturing and quality controls. Its configuration usually does not conform to any requirements and its purpose is to verify that the design meets performance requirements of the detail specification, Part I.)

Design acceptability depends on meeting the technical requirements of Part I of the equipment specification. Special emphasis is placed on reviewing detailed engineering and supporting analyses, interface considerations, and appraising the tolerances and performance parameters specified for the equipment. Approval of the design by the review board indicates that the equipment is ready for fabrication.

A list of requests for waivers[4] or deviations to the specifications is prepared a few weeks in advance of the CDR and submitted to the customer for review. To support this list of deviations, specification change notices may be used to describe each change and to obtain customer approval during the CDR. Only one change should be added to each SCN. Also, a list of the items to be reviewed at the CDR should be sent to the customer for review 30 days before the scheduled meeting.

14.7   FIRST ARTICLE CONFIGURATION INSPECTION (FACI)/PHYSICAL CONFIGURATION AUDIT (PCA)[5]

FACI is the last formal review held with the customer within the contractor's facility, excluding special situations where major changes are made to the approved configuration. The purpose of this review is to establish the final performance, mechanical, and electrical configurations of the equipment before beginning flight production. This process results in a product configuration baseline from which all future changes are referenced. This review is conducted at the completion of the first production equipment, and satisfactory completion of the inspection/audit results in the customer's approval of Part II of the detail specification for the equipment. Part II of the specification is delivered to the customer for review about 30 days before the review meeting is held.

The review includes as a part of its agenda the following areas: (a) an audit is conducted to verify the exact relationship of the equipment to its released

---

[4] If prototype fabrication was allowed to begin before CDR.

[5] FACI is AFSCM 375-1 terminology. PCA is DOD standard terminology recently established by DOD Directive 5010.19.

engineering data by comparing the equipment to its data; (b) test methods and data are reviewed and compared to the performance requirements of the equipment; and (c) determination is made of the exact relationship between the configuration of the equipment to be subsequently delivered and the configuration of the equipment qualified by the review. The third item is necessary because some changes from the first article may be desired in the next production equipments.

A major result of the review is a configuration baseline index and configuration verification record that identifies all key drawings, specifications, and test procedures to which the first production equipment was built. Any changes to future equipment can then be easily compared to this baseline document. Refer to Chapter 19 for additional information on the baseline index.

Requirements for this review differ to some extent from the previous contractual reviews. These differences are listed as additional functions performed during the meeting:

1. Verify that part numbers on drawings, parts lists, manufacturing orders or instructions, and the various pieces of hardware that make up the equipment agree.

2. Verify that hardware and manufacturing documents agree with special notes supplied by engineering drawings.

3. Verify that manufacturing operations are in the correct sequence.

4. Verify that the finish, color coding, identification, and workmanship are to approved specifications.

5. Verify that the hardware clearances are adequate and prevent interference among parts.

6. Verify that inspection stamps on manufacturing instructions are the correct type and that they are located properly.

7. Verify that serialization requirements are satisfied.

8. Verify that component types, part quantities, and location are correct.

9. Verify that change letters (see Chapter 12) on manufacturing documents, engineering data, and drawings agree.

10. Verify that drawings conform to drafting standards specified for the project.

11. Verify that the physical characteristics of the equipment conform to the drawing (dimensions, shape, finish, and so on).

12. Verify that packaging and cleanliness practices are adequate.

13. Verify that special handling instructions are correct and complete.

14. Identify time-limited items in CI.

15. Verify that nonstandard parts are identified and qualified for use in the CI.

16. Verify that approved deviations and waivers are available for all nonconformances to contract requirements.

17. Verify that all approved changes have been incorporated.

Documentation for this review is supplied to the customer about 30 days before the meeting and covers a broad range of subjects. Data items include drawings; test procedures; specifications; test and inspection data; failure and corrective action reports; lists of shortages, waivers and deviations; lists of parts that deteriorate with time; a record of all parts replaced during assembly and test; and a log of accumulated operating time for the equipment. A qualification status report (list) is also provided to identify the tests, processes, and data by which parts are certified as having met their specified requirements as to the level of performance and reliability required for the end use of the equipment. (See Chapter 7.13, *Data Package.*)

After the review is complete, a report is prepared by the project manager. This report covers the results of the review and includes each team captain's comments on the review and the status of each action item that resulted from the review along with its anticipated completion date. If, as a result of the meeting, a part or assembly of the equipment is unacceptable, the disposition of this hardware is given. The report also identifies the accepted configuration baseline established by the review inspection.

Note that customer notification to the company that the review has been satisfactorily completed does not free the company from complying with contract requirements.

Chapter 15

ENGINEERING
RELEASE RECORDS

Engineering release records are of paramount importance to the overall configuration management of an equipment. These records provide the official company data that identify and interrelate key engineering and administrative information describing the status of engineering documents, such as drawings, parts lists, specifications, manuals, and change documents (EO's, ECP's, SCN's, and so on).

Although certain minimum requirements apply, the types, format, and contents of these records depend on company and customer requirements. The records selected should be preprinted as shown in Figure 6.1 and 6.2 and should contain entries to accommodate government data requirements, although these entries need not be completed for commercial projects that do not require all these data.

To help guarantee the project manager and customer that the equipment conforms to its engineering data (initial plus subsequent releases), the four following minimum capabilities must be satisfied by the company to provide an effective engineering data release system:

1. Record and maintain key engineering and administrative data elements during the project and retain data for at least two years after the last equipment is delivered.

2. Release planning documents, design activity documentation, production documents, test data, and so on, to the project personnel.

3. Release engineering change documents to project personnel.

4. Release field changes to project personnel, including personnel at the deployment sites.

The following sections discuss the key areas related to engineering release records. Keep in mind that release records are primarily administrative documents that summarize who did what, who authorized it, what changes resulted, what was affected, and when it happened. It is not the purpose of these records to provide technical data.

15.1   RESPONSIBILITY FOR RELEASE RECORDS

The responsibility for maintaining release records lies with data control and the configuration manager. Data control is the key element in making the initial record entries and for updating the records as engineering data are changed by project personnel. The principal role of the configuration manager is to identify and define the release records required and to instruct and monitor the data control personnel in their preparation and maintenance of the records. Extreme care should be taken to ensure that only data control personnel are allowed to prepare or revise release records. Therefore these records should be kept in a restricted area that forbids other personnel from entering.

The ease of data retrieval is another aspect of release records that should be considered by the configuration manager who is responsible for selecting the record types and formats. Engineering personnel will rely on data control to provide them with instantaneous data on the latest revisions to a drawing or the new number for a part that has been superseded with a new design. For large programs, microfilm techniques, visual displays, and magnetic tape readouts containing the release data may be used.

In summary, the configuration manager and data control personnel are responsible for the following major tasks:

1. Establish release records at beginning of project and continually maintain them.
2. Keep accurate and complete records.
3. Provide data review procedures and verification requirements to assure correct serial number effectivity of design and correct incorporation of engineering changes into the manufactured equipment.
4. Replace controlled drawings whenever changes in equipment configuration occur.
5. Provide historical records for all actions that occurred from the beginning of the project.
6. Apply same release system to all projects regardless of the design complexity or number of design changes.
7. Prevent unauthorized changes to drawings.
8. Identify persons withdrawing documents for revisions.
9. Distribute documents to staff.

15.2   CONTENTS OF RELEASE RECORDS

Engineering release records contain key information on the change status, date of release, distribution, and so on, on each document released by engineering. Data control is responsible for maintaining these records in a central file area. The records—status cards and release and issue forms—are always available for inspection by the customer, quality assurance representatives, or the project office. When engineering orders or other change documents are officially approved and released through data control, the release records are updated to identify the EO title, number, and date of issue. When a vellum is released to engineering for revision, the records are changed to indicate the name of the person revising the vellum and the date that it was released to him.

Engineering release records can contain up to 25 data elements; however, at a minimum these records should completely and uniquely identify the document, the release authorization, and the approved changes to the document. Specifically the records should contain the data given in Table X. Secondary but important additional data elements include the following:

1. Drawing size.
2. Number of sheets to the document.
3. Date that engineering anticipated release of the document.
4. Manufacturer's code number.
5. Quantity of items.
6. Top drawing number.
7. CI specification number.
8. Type of drawing (assembly, detail, installation, source control, or schematic).

TABLE X
Minimum Release Record Data Requirements

1. The name, number, and type of document
2. The CI or equipment number
3. The project title and contract number
4. The project work authorization number issued by the company
5. The originator of the document
6. For drawings, the next higher assembly number
7. Document change letters and part number suffixes
8. Change document titles and numbers
9. Equipment serial number effectivities of each change
10. Type of release: production, reference, advance, limited, and so on
11. The signature of the data control clerk who issued the document to the staff
12. The signature of the project manager or configuration manager who authorized the document for release by data control
13. The distribution code or persons who received copies of the document

15.3 DUAL RELEASE OF DOCUMENTS

Only one release record and source should be allowed for each drawing, part list, specification, and so on. Although different types of documents can be released by separate control groups, it is recommended that all types of documents be released from one central data control group. Drawings released by a subcontractor should not be re-released by a contractor or other agency if the problem of discrepant drawings is to be avoided. For example, a contractor that releases a subcontractor drawing to a vendor for manufacture or supply of the item depicted on the drawing may find a few weeks or months later that the subcontractor has released a revised drawing that has made the new items noninterchangeable with previously built items. The result may be additional costs and delivery slippage due to the rework effort that will be required to update the item. Therefore, the latest released make-to (production) drawing should be obtained directly from the responsible design group with a request to know whether any design changes are in process or planned. With this information, the contractor can determine whether it is possible to wait for the revised production drawing before issuing a purchase order for the item.

15.4 MULTIPLE RELEASE OF DOCUMENTS

Engineering data released after the production design is committed for the equipment are processed by using a multiple release system. This means that data for the old and new designs are kept in the data control active data files. The old data are stamped "Reference Only" and "MRR."[1] Of course, the make-to drawings are stamped "Production" as described in Chapter 6.2. The status record (see Figure 6.2) should identify the old document entry as follows:

Date Released	Change Notice Number	Level	Distribution	Remarks	Change Letter
8/20/71	—	2	D	Prototype (MRR)	N/C
9/3/71	EO 001	1	D	Flight	N/C-1

Note that the "MRR" designation indicates multiple or concurrent release with all the subsequent entries unless a line is drawn through "MRR." The above example is given only for illustration and does not prescribe the system that should necessarily be used.

When an EO is attached to the original drawing, the EO and drawing constitute a multiple release where the drawing effectivity applies to all equipment serial numbers except as amended by the effectivity data given

[1] MRR represents multiple reference release.

on the EO. Thus a no change drawing may apply to equipment serial numbers 1 through 10, but the EO changes the effectivity for the engineering change described for equipment serial numbers 5 through 10. Therefore the drawing originally released and the EO provide both configurations in one package.

The multiple release system is retained for the applicable document unless all of the delivered equipments are retrofitted to incorporate the latest change. Accordingly, the effectivity of the equipment would return to "1 and subs." Upon retrofit, a change verification report is sent to data control indicating that all equipments have been upgraded to the latest design. A data control clerk then removes the superseded document from the active file, stamps it "Superseded," and places it in the history file. If the MRR system using EO's is being used, no change is made except for crossing out the MRR on the drawing status card. Of course, if the vellum is revised to incorporate the EO, the revised drawing will be the only authorized production drawing available in the active file.

15.5   DATA CONTROL RELEASE FUNCTION CAPABILITIES

To satisfy government requirements and to provide engineering personnel with required data, the data control release system should be capable of providing data on parts used, effectivities, parts quantities, change documents, and standard parts for the equipment. Using the release records and the engineering documentation, that is, drawings, parts lists, source and specification control drawings, the following data (usually accumulated by mechanized means) should be retrievable upon request of the customer or project member; see table on page 206.

15.6   CHANGE DOCUMENTATION

Drawing status and change document lists (see Chapter 17 for a description of these documents) are also maintained by data control, or the group responsible for releasing the data, to provide current data on change status. These lists identify all EO's and ECP's in process or released against the equipment. The change document numbers are identified and the applicable drawing or part number specified. The change class, I or II, is also recorded. If a release and issue form is prepared for each change document, the effectivity of the change is recorded on the form as shown in Figure 6.1.

15.7   IDENTIFICATION OF ENGINEERING CHANGES

Data control must be capable of identifying all engineering changes and of retaining records on superseded configuration requirements affecting equipments formally accepted by the customer.

Data Element	Source
1. Composition of any part number at any level in terms of subordinate parts.	Parts lists
2. Next higher level assembly part number for any part.	Drawing/Indentured parts list
3. Composition of the CI in terms of subordinate assembly numbers.	Top assembly drawing/ indentured parts list
4. CI serial numbers (effectivity) on which any subordinate part is used.	Release and issue forms, change documents, VCR
5. Change identification numbers released against the CI.	ECP/EO lists, status cards
6. CI numbers and CI serial numbers that constitute the effectivity of any change identification number.	Release and issue forms, change documents
7. Company standard part numbers used in a nonstandard part.	Drawings, parts lists, non-standard parts list for CI
8. Vendor part numbers issued as a result of special requirements.	Specifications, source or specification control drawings
9. Quantity of parts per assembly and CI.	Parts lists and material control records

All Class I and Class II engineering changes released for production incorporation are identified by the change identification numbers and must be completely released before formal acceptance of the equipment by the customer. The configuration identification data and descriptions released for each equipment at the time of its formal acceptance are retained in release records for the time required by the contract or by company policy, usually two years.

All engineering design changes released for incorporation in any equipment that has been formally accepted by the customer must be identified by unique company change identification numbers.

15.8 FIELD CHANGE IDENTIFICATION AND RELEASE

Engineering data defining formally accepted equipment that is in field testing must be maintained current with all field activity design requirements and released as described for the following three conditions.

Before FACI. Sometimes a prototype or production equipment is delivered to the customer before FACI for field testing or field operation. If design changes are made to the equipment during field work, a design change package completely describing the change must be returned to the company. The drawings, parts list, and wire list are then revised and officially released for production by data control. If no other equipments have been delivered, the old or superseded data are removed from the active files and placed in the historical file in data control. If the responsible field engineer has not checked the revised documents before release, he should verify that they are correct when he receives them. Any discrepancies should be immediately reported to the project engineer at the company facility for correction.

After FACI. After FACI, superseded requirements data must be retained for reference, and superseding requirements data, resulting from field work, are added to the release records for all equipments that have been formally accepted by the customer after FACI and are under control of the company. Effectivities for superseded parts coded as a requirements release (manufacture or procurement required) are retained in the records along with the effectivities for superseding parts coded as a reference release (manufacture or procurement not required). The serial number effectivities for the reference and requirements releases should be recorded on the drawings and the status card. The sum of the limited effectivities for old and new parts equals the original effectivity. Superseded requirements data are retained in multiple release records until status accounting records indicate superseded configurations are no longer in inventory; that is, all equipments have been reworked to the latest design configuration.

At Customer's Facility. Engineering change data for equipments formally accepted by the customer which are no longer the responsibility of the company are released for customer equipment retrofit action at his facility or site. It is important that the new documentation be as complete and accurately prepared as during the project.

A request for equipment retrofit modification may arise as a result of customer request, or revised contractor performance, quality, or reliability requirements. The specific requirement for authorization of any equipment modification is the release of an approved engineering order and a contract change notice or other contract authorization. It is the responsibility of the contractor's logistics or product support group in the field (field engineers) or the customer, whichever is called for in the amended contract, to incorporate changes in the affected hardware. The incorporation of the change is accomplished by the installation of a "retrofit modification kit" consisting of the physical parts and materials to be installed, modification kit drawings and parts lists, reidentification labels or plates (or in some cases stenciling), and modification instructions.

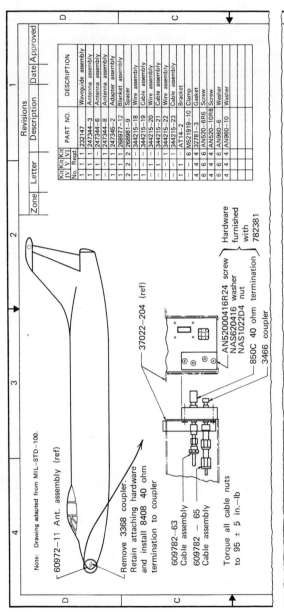
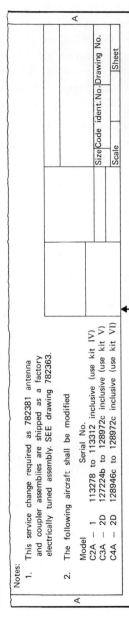

Figure 15.1a Modification kit drawing.

Name of Complete Article to be Modified:		ECP Number:	Used On:	Part Number:

Name of Part or Lowest Modified Article: Other: Part Number:

Nature of Modification:

Modification Requested by:
□ ABC   □ Customer

Reason for Modification:

Recommended Priority:
□ Emergency
□ Urgent
□ Routine

Items Affected by Modification:

□ Performance   □ Interchangeability   □ Contract Price
□ Government Furnished Equipment   □ Service Life   □ Overhaul Methods
□ Safety   □ Aerospace Ground Equipment   □ Gov't Furnished Property   □ Spare Parts Contract
□ Contract Weight   □ Delivery Schedule   □ Ground Support Equipment   □ Other _____
□ Maintenance Procedure   □ Operating Procedures   □ Special Test Equipment   □ Other _____

Publication Affected by Change:

Task	Skill	Man/Hours Required

Total Man/Hours

Remarks:

	Code Identification Number	Size		Rev.
(ABC) Corporation, Washington, DC	09876	A		Sheet

Figure 15.1b Modification summary sheet.

In order to continue proper identification requirements, the modification kit receives a control number as shown by this example:

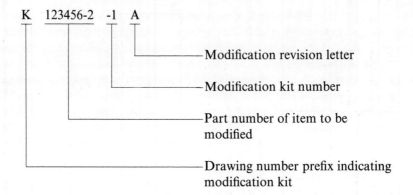

K   123456-2 -1 A

Modification revision letter

Modification kit number

Part number of item to be modified

Drawing number prefix indicating modification kit

The contractor's product support group, or field engineers, responsible for releasing retrofit modification design packages must also assume responsibility for (a) "exercising" the performance of the retrofit kit before release of kit drawings for procurement and fabrication; (b) incorporating and "field recording"[2] retrofit modifications into equipment, except where the customer assumes this action; and (c) submitting retrofit modification incorporation information to quality assurance for maintenance of configuration records.

It should be noted that quality assurance provides quality control for all retrofit modification kits and witnesses and approves all retrofit modification kit trial runs before release of the drawing for procurement and fabrication.

Retrofit Modification Kit Drawing Package. The retrofit modification kit drawing package is a collection of data that provides all instructions, manhours and skills information, identification of parts and materials, identification requirements, and other information required for the performance of an equipment modification. Typical elements of the kit drawing package are illustrated in Figure 15.1.

The release status record for the affected drawing should be updated to identify the field changes made.

---

[2] Field recording may require formal time compliance technical order (TCTO) activities or may be simply an entry into the equipment log for the item being modified.

Chapter **16**

VERIFICATION OF CHANGE INCORPORATION

No matter how carefully project personnel prepare and review engineering change documents (EO's, ECP's, and so on), these documents are worthless if the changes are not made to production equipment and properly identified in release and manufacturing records. Therefore a principal configuration identification task is to verify that engineering changes are incorporated as described in revised engineering documents and in accordance with customer contract requirements.

Because change incorporation involves all levels of project personnel, explicit policies issued by top company management, detailed procedures, and complete records are necessary to assure consistently good results. These procedures must be written and published to satisfy the customer if he decides to inspect the company's change verification system.

Change verification requires a systematic approach for assuring that:

1. All Class I changes have been approved by the customer.
2. Released engineering data and purchase orders meet contract requirements.
3. Changed items are built and installed into the equipment as specified in released engineering data.
4. Engineering change incorporation actions are documented as required for formal acceptance of the equipment.

16.1 VERIFICATION REQUIREMENTS

The quality assurance engineer and manufacturing engineer are responsible for verifying that the serial number effectivities of engineering changes agree with CCB and CPG actions that approved and authorized the change. Effectivity data are always specified on the ECP and EO, which

TABLE XI
Configuration Verification Responsibilities

Area	Responsible Person	Verification Mode
Release of SCN or specification as approved by SCN.	Configuration manager Project manager	Check specification index listing SCN's and their ECP's.
Alteration of qualification requirements, such as environmental tests, must be authorized by a contract change.	Configuration manager Project manager Project engineer Test engineer Quality assurance engineer	Check contract change notice or customer approved ECP for authorization. Also check waivers and deviations.
Acceptance test procedure is revised to agree with changed engineering data without conflicting with contract requirements.[a]	Project manager Project engineer Test engineer Quality assurance engineer	Check contract revision, contract change notice, ECP's, and EO's.
Vendor purchase orders and change notices agree with contract requirements.[a]	Procurement representative Quality assurance engineer Reliability engineer Project manager Project engineer	Check approved ECP's, EO's, specifications, and test procedures.
Manufacturing requirements agree with contract requirements.[a]	Manufacturing engineer Quality assurance engineer Reliability engineer	Check applicable ECP's, EO's, and specifications.

[a] Customer-approved specifications, effectivities, schedules, and costs.

are the source documents for effectivity and other change data. However, release and issue records, production drawings with EO's attached, and manufacturing orders also contain effectivity data for each change. The as-built list described in Chapter 19 is also called the verified configuration record. This record is prepared by the configuration manager and quality assurance engineer who inspect the equipment and manufacturing orders to determine the drawing number, its revision letter, and the EO numbers to which a serial numbered item was built. Discrepancies between this list and contract requirements must be resolved by the project manager and customer representative. If the discrepancies are in violation of the contract, waivers may be prepared allowing the equipment to be delivered to the customer in its present configuration. The waivers must be signed by the customer's technical officer or contracts administrator, or both, depending on contract requirements.

The manufacturing order is the principal document by which the manufacturing engineer controls the fabrication of the equipment, the incorporation of changes, and the identification of released engineering data. The manufacturing order (MO) is a translation of an engineering drawing into a step-by-step procedure for building or assembling an item. One MO, identified with a unique internal company control number, is prepared for each part number and each serial number. In addition to manufacturing instructions, the MO includes inspection points that occur at critical points in the procedure. Upon satisfactory inspection of the partially completed item, the quality assurance engineer places an approval stamp next to inspection instruction entry on the MO, thereby authorizing continuation of manufacturing operations.

When an EO is released by data control, the manufacturing engineer receives a copy stamped "Production." Using this copy only, he prepares a revised MO or adds a revision sheet to the original MO for each serial numbered item affected by the EO. The EO number and revised drawing number or letter are recorded and provide the as-built record for each item built by the company. The revision sheet is identified with the original MO number and a revision letter. The EO number is also added to the original MO in a block provided on the form.

Other key areas affected by a change are listed in Table XI. The responsible project members and mode of verification are also given.

16.2   VENDOR ITEM REQUIREMENTS

The manufacture of vendor items to the latest released engineering data is as important as company compliance to new requirements. Contractually, the company is usually responsible for assuring that its subcontractors, suppliers, and vendors meet all requirements for verifying that Class I changes have been incorporated. Therefore the material control engineer must be capable of identifying each engineering change that has been made to all vendor items by serial and part number. To do this, the purchasing representative must prepare the purchase order so that he imposes on the vendor all the requirements that apply to the company. In addition, the configuration manager and quality assurance engineer should verify that the vendor is following satisfactory procedures for assuring the company that engineering changes are being made to items as required by applicable company or vendor ECP's or EO's.

Another area of control applying to vendor items containing changes is their proper routing to the correct serial numbered equipments (CI's) that are designated to include the changed items. To accomplish this routing, the CI serial number effectivity of the items should be specified by the vendor in his shipping document for each item, based on the effectivity supplied to the vendor when a PO change notice is issued as a result of an engineering change. Thus, after receiving inspection is made at the company, the item and its effectivity data can be delivered to the material control area for assembly into a parts kit that will be issued to the same serial numbered equipment.

Purchase order requirements should specify that engineering changes initiated by a vendor or subcontractor must be identified as Class I or Class II changes and submitted to the company through the contracts administrator. Vendor Class I changes should require company approval before they are incorporated.

16.3 INSPECTION OF EQUIPMENT CHANGES

To assure accurate and complete change incorporation into production equipment, the quality assurance engineer must inspect manufacturing operations on a daily basis, audit applicable documents for completeness, accuracy, and currency, and verify that standard shop and documentation practices are being followed by manufacturing personnel. In particular, the quality assurance engineer reviews manufacturing orders to verify complete and accurate translation of engineering data for each engineering change into step-by-step manufacturing instructions. He checks each operation for completeness and agreement with the engineering data as well as with contract requirements for soldering, welding, encapsulating, and other workmanship standards. To provide independent evaluation of manufacturing data, the quality assurance engineer uses his own set of drawings, parts lists, and so on, to make sure that manufacturing is using the latest revised and released data. He can also check with data control to verify the accuracy and currency of his own data if he lacks confidence in the company's release and distribution system.

The first installation of a change on an item requires special attention because personnel must change their established operating modes and procedures to accommodate the engineering change. Therefore the quality assurance engineer must take extra care to inspect material control for correct parts kitting, production control for revised MO's, and assembly operations for accurate incorporation of the change to ensure its proper installation as shown in the engineering data. If for some reason—for example, missing parts—manufacturing cannot incorporate the change, the quality assurance engineer verifies that a shortage report is completed by manufacturing and that it accompanies the item through final assembly, when shortages for all items will be compiled by the quality assurance engineer into one list or group as described in Chapter 7.13.

Another important item to be checked is the sequence of ECP or EO incorporation into the CI. Many times changes must be made in the exact sequence indicated by the ECP numbers. When this situation occurs, the sequence of incorporation should be given on the EO to keep manufacturing from making an out-of-sequence change when preceding ECP's and EO's have not been approved for incorporation while subsequent change documents have. The quality assurance engineer should be particularly careful about monitoring this situation and should work closely with the project engineer, configuration manager, and manufacturing engineer to avoid an incorrect change incorporation sequence which could cause delays due to rework, damage to parts, and additional costs.

In addition to the previous inspections, the quality assurance engineer verifies that the requirements for subsequent incorporation of each change are given in production orders or MO's and that routine material and manufacturing controls are adequate to assure the building of items to the changed requirements or the preparation of shortage reports when the changes cannot be made. Materials control should be checked to ensure kit preparation using authorized parts, complete and accurate part identification, removal of obsolete parts, and release of kits only to the authorized manufacturing engineer for use in specific serial numbered equipments. Manufacturing controls should be inspected to verify that the kit list agrees with parts and serial numbers, that items with apparent defects are replaced with good ones, and that parts are assembled in accordance with the latest authorized manufacturing instruction and engineering data.

A final quality assurance function is to maintain an adequate surveillance of all operations to ensure that approved standard practices are being followed to incorporate and document subsequent installations of each engineering change. Surveillance practices include periodic audits of manufacturing orders for instructions describing the change to be made, inspection of the item to verify that the change has been made as described, and a check to ensure that the change effectivity corresponds with the customer's CCN, ECP, and EO.

16.4 DOCUMENTATION OF INCORPORATED ENGINEERING CHANGES

The company is responsible for maintaining engineering change incorporation records. Usually manufacturing and quality assurance are the primary areas for this effort, with the configuration manager verifying the adequacy and proper functioning of the system. Incorporation records include (a) an inspection file showing evidence of inspection acceptance of each equipment in its required configuration and (b) a record showing evidence of inspection verification of the complete incorporation of each engineering change in the first equipment for which the change is effective. Remember that part number identification alone does not verify that a change has been incorporated because the level of assembly where an engineering change has been installed and related to a particular equipment may be below the level of assembly where interchangeability was reestablished. Thus both the serial number and part number identifiers are required.

An inspection record is required for each equipment previously inspected and accepted by quality assurance when changes to the equipment result in the removal, replacement, or rework of equipment parts after initial assembly. This record includes a description of the changes made, by whom, when, where, the authorization for the changes, and the tests or checks performed upon completion of the changes. An inspection acceptance stamp may also be required for the change. This stamp is placed next to an inspection operation on a revised or amended MO or in the equipment log book if the equipment is completely assembled.

A shortage report is also required for each equipment to record any part of an engineering change that has not been installed or completed before delivery to the customer. This report should identify the planned incorporation date and location where changes will be made to the equipment. (The shortage report is basically a list and not a formal report.)

16.5   CUSTOMER ACCEPTANCE

Customer acceptance of engineering changes in manufactured equipments is accomplished upon formal acceptance of the equipment at the company's facilities or where the change is made. The customer bases his acceptance upon visual inspection of the equipment when changes can be readily seen or upon documentation which verifies that the changes have been witnessed and approved by the quality assurance engineer or his representative. The quality assurance buy-off stamp on the MO is an example of the documentation that is used to verify inspection of change incorporation. In particular, formal customer acceptance is based on the following criteria:

1. Contract terms and conditions, including contract modifications, waivers, ECP's, and deviations.

2. Identification requirements (see Chapter 8), such as identification plates, serial numbering of components, and part numbering of all items.

3. Written agreements reached at previous configuration management reviews.

4. Agreement of contract change notices, company inspection acceptance files, change incorporation records, breaks of inspections,[1] waivers, deviations, and shortage records at time of equipment acceptance.

---

[1] A break of inspection (BOI) refers to a condition that occurs when rework, removal, or replacement work is performed on an equipment already inspected and approved by quality assurance or the customer's quality assurance representative.

PART III   CONFIGURATION ACCOUNTING

Configuration accounting is the payoff for the configuration management system which assures the customer that he is getting what he pays for. The accounting aspect of configuration management provides the customer with the documentation that describes the status of the CI design and the final configuration of each delivered CI. The effectiveness of the accounting function is dependent on the quality of the identification and control efforts. For example, in financial accounting the work order structure and cost control disciplines are vital in providing meaningful accounting information. Accounting can be a routine task or a nightmare, depending on the identification and control aspects of the configuration management system used for the project.

To implement configuration or baseline control procedures described in the previous chapters, all personnel responsible for supporting the CI acquisition phase must cooperate to meet the requirements established for each baseline or for changes to the baselines. To achieve this objective, the exact definition of the baseline and all changes made to it must be available to project and customer personnel. In addition, the actions of project personnel in response to approved baseline changes must be reported. This reporting enables management to determine the status of approved changes and to take actions to ensure that directives and decisions are being carried out.

Baseline identification and control, change monitoring, and management follow-up control of directives are achieved through the following:

1. Maintain a configuration record documenting all approved changes to the CI.

2. Implement configuration identification and status accounting of a CI mission, design, series when the product configuration baseline is

approved. (Reports will be maintained until responsibility for the product is transferred to the customer.)

3. Identify, accomplish, report, and account for all changes during installation and checkout at the system site and to inservice CI's.

4. Report actual location and status of each CI during acquisition phase.

In major space or other types of systems, the number of CI serial numbers and the number of changes to CI's can run into thousands. To control and relate these changes, we need to know more about a change than that it was authorized or accomplished. Therefore the following information is desirable for a configuration accounting report:

1. Date the change was identified.
2. Who identified the change.
3. Reason for the change.
4. Urgency of the change.
5. Description of the change.
6. Scheduled submittal date.
7. Actual submittal date.
8. Required decision date.
9. Actual decision date.
10. Required authorization date.
11. Actual authorization date.
12. The decision made (change approved or rejected).
13. Authorization document.
14. Scheduled engineering release.
15. Actual engineering release.
16. Scheduled procurement.
17. Actual procurement.
18. Receipt of material.
19. Scheduled tool fabrication.
20. Actual tool fabrication.
21. Scheduled part fabrication.
22. Actual part fabrication.
23. Scheduled assembly.
24. Actual assembly.
25. Scheduled installation.
26. Actual installation.
27. Place of installation.
28. Who is responsible for installation.
29. Man-hours required to install.
30. Power on or off condition during installation.
31. System down time during installation.

32. Spare parts to be reworked.
33. New spare parts required.
34. Technical manual needed.
35. Technical manual to be changed.
36. Estimated cost of change.
37. Actual cost of change.

The management process consists of three steps:

1. Make a decision.
2. Tell everyone who needs to know what the decision is.
3. Follow-up to verify that the decision is carried out.

Configuration accounting is the management tool that the configuration management office uses to implement steps 2 and 3. It is the purpose of Chapters 17 through 19 to describe configuration accounting documents and procedures. However, keep in mind that the information presented is only representative and does not cover the field completely.[1]

---

[1] This section was adapted from notes supplied by Walker Bennett, Nov. 1969.

Chapter 17

DRAWING AND ANCILLARY RECORDS

Drawing and ancillary records are fundamental to configuration accounting. They identify all drawings prepared for the product and their current status; for example, released to staff, in revision, revised, or obsolete. The key records maintained by the configuration manager or data control are the following:

1. Drawing status record.
2. Drawing lists.
3. Drawing and engineering order lists.
4. Engineering order and ECP log books.
5. Document release records.
6. Document sign-out cards for vellums released back to engineering for changes.
7. A sketch log book.
8. A Rolodex cross-reference file for interrelating change documents to their drawings or authorization notices.

The format and content of these records are described in the following sections. All these records may not be required for a project and their use will depend on company or contract requirements.

17.1 DRAWING STATUS RECORDS

Drawing status records (see Figure 6.2) are usually preprinted cards (for example, five by eight inches) with entries for all major data concerning a drawing identified on the card. Before reproduction of the document, a card is filled out by data control personnel for each drawing, parts list, wire list, specification, and so on, when the document is released to data control for distribution to the project staff or the customer. Typical entries

(ABC) Corporation Washington, D.C. Project _____			Drawing Log	
Size	Drawing Number	Description	Date and Draftsman	
A	123000	Power converter assembly	4/15/70 Stontz	
A	123001	Envelope drawing	4/18/70 Stontz	
B	123002	Block diagram	4/20/70 Stontz	
C	123003	Chassis	4/22/70 Hill	
A	123004	Circuit board	4/27/70 Hill	
A	123005	Circuit board master	5/1/70 West	

Figure 17.1   Sample drawing log.

include the following:

1. Project or job number.
2. Document identification number and revision letter.
3. Title of document.
4. Size of document (A, B, C, D, E, or roll. These letters represent specific document dimensions commonly used in the industry. A is the smallest size at 8 1/2 by 11 inches, and a roll size is 34 inches wide and can be almost any length because it is cut from a roll to meet drawing requirements).
5. Assembly number that the part shown on the drawing is used on.
6. The CI number that the part or item is used on.
7. Date the document was released by engineering to data control.
8. Date that the document was distributed to the staff.
9. The type of release: production, reference, advance, or limited.
10. Number of sheets in the document.
11. Special comments or remarks, such as the type of release, can be entered under this heading if the type of release is not specifically given a separate heading on the card.

Each time a change is made to the document, the authorizing EO or SCN number is recorded in the "Change Notice Number" column. The date of the released drawing or change document is also recorded. When a document vellum is withdrawn, the person issued the vellum signs his name in the "Remarks" column. (A separate sign-out card may be used also.) Entries are made on the card when a part or drawing is made obsolete by a newly numbered part or drawing as shown in Figure 6.2.

A separate drawing log is also kept listing all released drawing numbers (see Figure 17.1). This log supplements the status record cards and provides a convenient historical record of all the drawings issued on the project. It is prepared by the design engineer or lead draftsman who submits the log to data control upon completion of the project. It provides a convenient, official record of all the project drawings in a readily available form that does not require much filing space.

17.2 DRAWING LIST

Drawing lists are prepared on a regular basis to maintain the current status of the drawings for the equipment. The list is prepared every week or two, depending on the project staff requirements or company policy. It lists all released drawings in consecutive or numerical order and identifies the completion or current change status for each drawing. (If parts lists are separate from the drawings, these are also listed.) The list includes the full title or a shortened version of it for each document. Updating of the list

Drawing and Engineering Order Status List

Drawing Description	Drawing Number	Drawing Revision	Drawing Status	Release Date	Engineering Order Status	
					Attached EO's	In Process EO's
Amplifier, pulse	123456	A	Released	7/3/71	0002, 0004, 0008, 0012	0030, 0035

Figure 17.2　Drawing and EO status list.

is done by data control personnel who mark up a master copy as the drawing status record file is checked card by card for new drawings or EO revisions to drawings. However, it should be examined by the configuration manager or the designer before it is released.

17.3   DRAWING AND EO STATUS LIST

The drawing and engineering order (EO) status list is an index of the latest EO's written against a drawing or parts list. This list provides a ready reference for determining whether the drawing is complete or in preparation and whether EO's have been released by engineering and attached to a drawing by data control, or are in the process of being approved by the project office or the customer. An illustration of the drawing and EO status list is given in Figure 17.2. Note that two key blocks apply to the EO status column:

1. Numbers of EO's attached to drawing, parts list, and so on.
2. Numbers of EO's in process.

An in-process EO is one that has received the project manager's approval to be submitted to the configuration control board but has not been formally signed off and released for attachment to the drawing. The revision column applies to the letter change to the drawing; the "Release Date" column is used to record the date that drawing was released by engineering for manufacturing.

17.4   EO LOG BOOK

An engineering order log book is maintained by the configuration manager. This log book is used to issue engineering order numbers and to ensure that the descriptions and names of requesters are recorded for issuance to the project staff. Each number should be issued in sequence only. The following data should be recorded for each EO:

1. EO number.
2. Date of issue.
3. Requester's name and department.
4. EO title or change description.
5. Drawing number and revision letter affected.
6. Change class, I or II.

17.5   RELEASE FORM RECORDS

The release and issue forms described in Chapter 6 (see Figure 6.1) are kept on file as formal records of engineering release of a document to data

	ABC Corporation		
	Washington, D.C.	**Sketch Log**/1400-SK-X-XX	
	Project_____		

Size	Sketch Number	Description	Originator
A	1400-SK-V-01	Dual latch driver	R. Villot
B	1400-SK-V-02	FET switch	''
A	1400-SK-V-03	Current detector	''
A	1400-SK-V-04	Operational amplifier	''
A	1400-SK-V-05	One shot	''

Figure 17.3   Sample sketch log.

control and the date of issue to the project staff. They also identify the originator of the document, authority for release, and who received copies of the document. These records can be stored in a separate five- by eight-inch card file in numerical order until needed to verify the status of the document or to obtain other related data needed by the configuration manager or project manager. A new release record is prepared for each document or revision released, including EO's. The most current card is maintained in the active file and the previous cards are filed in a history section. Data control personnel use these forms to complete and update the drawing status cards described in section 17.1.

17.6 SKETCH LOG BOOK

Sketches produced before formal configuration control begins should be retained for reference by data control. Each sketch is numbered and a copy stored in a data control project sketch file. A sketch log book is also kept. This book (see sample page in Figure 17.3) is used to record all sketches submitted by engineering to data control. Before accepting the sketch, data control personnel check it for correct identification. A simple sketch numbering system is given:

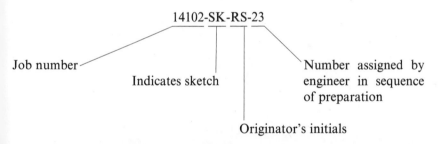

Data added to the log book (see Figure 17.3) include the sketch title, identifier, and originator. Normally, no distribution is required. However, if the engineer requests distribution, the persons specified are issued copies by data control.

17.7 SIGN-OUT CARDS

Sign-out cards are used in many companies to record the date and name of the person withdrawing a document vellum from data control. This card can be used instead of the status card for tracing the location of a missing vellum and for notifying people requesting document copies that a revision is in process. The card is 9 1/2 by 12 inches (other sizes are also used) and is kept with each drawing file folder when the vellum is withdrawn. The advantage of the sign-out card is that the data control personnel are flagged of a change in process when a staff member asks for a drawing copy. The inquirer can thus be notified that the drawing is in revision. The card, of course, contains the document number and revision next to the signature of the person who was issued the vellum. Sign-out cards can also be kept with the status cards described in Chapter 17.1.

17.8   ROLODEX FILE

A Rolodex file (see Figure 17.4) is a set of small cards (e.g., 2 1/2 by 4 inches) mounted on a spindle that has a knob for rotating the cards. This file provides a convenient, fast method for obtaining change status data related

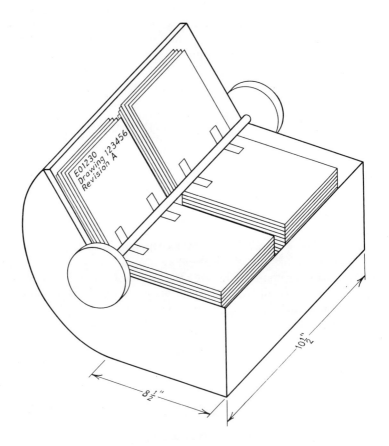

Figure 17.4   Rolodex file.

to a drawing. These cards contain EO's, SCN's, ADCN's, and part numbers that can be used for cross-referencing to find corresponding drawings or to identify all EO's issued on a drawing. The Rolodex file is maintained by data control.

17.9   CUSTOMER CHANGE APPROVAL RECORDS

All customer correspondence, teletypewriter exchange messages, contract change notices, and so forth, that apply to document changes are kept in a project file in data control. Since these documents represent the company's authorization by the customer for spending additional money or making technical changes to the equipment, it is imperative that a complete and accurate file be maintained and controlled.

17.10   OBSOLETE DISTRIBUTION LISTS

Copies of old distribution lists should be retained by data control. Occasionally questions arise as to who received copies of certain documents issued months or perhaps years ago. These records, along with the release form, can provide the required data. A separate folder for the distribution lists can be maintained in the obsolete drawings file and should be the first folder in the file, clearly labeled. All out-dated lists should be stamped "Obsolete."

17.11   OBSOLETE DRAWINGS FILE

A vellum (brownline) file of all obsolete drawings and other documents is maintained by data control. This file provides a historical record of the configuration of the component, part, subassembly, or equipment from the beginning of the project design effort. This record is a useful reference to engineering and is also required by the government for flight quality production equipment contracts. Note that a brownline or Xerox vellum is made from the original document before it is revised to incorporate engineering changes. It is best to make a reproducible copy of the original upon its release to data control for distribution. This reproducible copy should be kept in the active file with the original until a revision is made to the original.

Upon release of the revised original, the reproducible copy is stamped "Superseded" and placed in the inactive or history file for possible future reference. No other copies need be retained.

Chapter 18

SPECIFICATION RECORDS

Although drawing and specification records overlap, the specification records have their own special form and use and are therefore presented separately. Specification records include the specification change notice (SCN) change log, specification identification index, system configuration chart, and end item (CI) configuration chart. The purpose and content of each record are described below.

Specification change notices also fall into the class of specification records but are not covered here because they were described in Chapter 5. Remember that SCN's are used to record exact and complete change information relative to *particular* changes to approved technical requirements documents, such as system, detail equipment, and critical component specifications.

18.1 SCN CHANGE LOG

The SCN change log is used with all types of specifications to record customer or contract authorized SCN's to an approved specification. Thus, the change log identifies all SCN's released against a specification. The log is distributed as a cover sheet to an approved SCN or group of SCN's.[1] The change log sheet is placed in front of the first section of the specification. Figure 18.1 shows a sample specification change log. The SCN numbers are added in sequence as they are prepared and approved. Normally an ECP is required by the customer before approving an SCN to a contract imposed specification. This number goes into the ECP number column. The SCN date and affected pages are also filled in. If no pages are affected, the word "log" is added to indicate that this SCN only adds the ECP to the specification and does not change the existing text. The "Item Affected" column receives the serial numbers of the CI's affected by the change. When an SCN

---

[1] Accumulation and submission of a group of SCN's instead of individual ones is sometimes permitted by the customer when feasible.

(ABC) Corporation Washington, D.C.	Code identification number 09876	Specification Change Log	Date _____ Superseding Date _____

Specification number ———— Equipment nomenclature ———— CI number ——

SCN Number	ECP Number	SCN Date	Pages Affected	Item Affected
1	1000	4-7-71	Log	S/N 01 and up
2	1004	5-10-71	3	S/N 04 and up
3	1010	5-15-71	20	S/N 05 and up
4	1050	8-24-71	3 and 4	S/N 05 and up

Figure 18.1   Specification change log.

is released against a system specification, the effectivity ("Item Affected" column) is not usually recorded because systems are not identified by serial numbers, instead N/A is recorded to indicate "Not Applicable."

18.2 SPECIFICATION IDENTIFICATION INDEX

The specification identification index is a record of all specifications and approved changes to these specifications. A separate index sheet is prepared for each specification by the integrating or systems engineering contractor, and the first issue of the index includes all specifications that are a part of the design requirements baseline. When new specifications are required, a new index sheet is prepared for each one and distributed.

Figure 18.2 shows a sample specification identification index. Most of the data are self-explanatory. However, note that a "C" is shown after block 1. This letter indicates that the submittal for Part II of the detail equipment (product) specification was completed on the date given. It is also very important for the configuration manager to identify all other specifications and the companies affected by the change and any other SCN's that relate to the SCN number identified in the block 2 column. The last column identifies the new numbers of the parts changed by the SCN.

ABC Corporation Washington, D.C.	Code Ident. 09876	Specification Identification Index			

Nomenclature	Specification Number	Date
		CI number
		Part number

FACI scheduled date: 2 April 1971

① Specification, Part II, scheduled submittal date: 2 Mar 71 C

Specification, Part II, scheduled approval date: 8 April 71

SCN	ECP	Title	Other Specifications Affected	Related SCN ②	Changed Part Number
1	1001	Calibration change	AE 66-035	35	123456-210
2					
3					
4					
5					
6					
7					
8					
9					
10					

② Enter SCN number for associate contractor specification affected by the change.

Figure 18.2 Specification identification index.

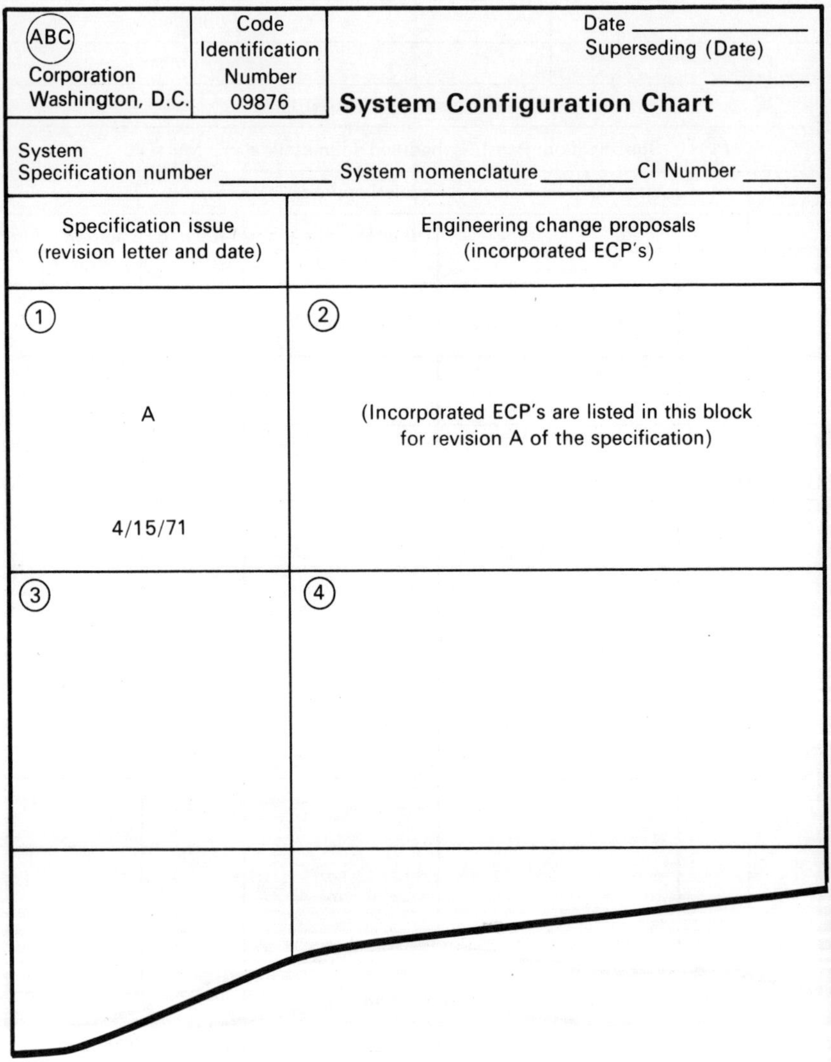

	Code Identification Number	Date _____
(ABC) Corporation Washington, D.C.	09876	Superseding (Date) _____
		System Configuration Chart

System
Specification number _____ System nomenclature _____ CI Number _____

Specification issue (revision letter and date)	Engineering change proposals (incorporated ECP's)
① A 4/15/71	② (Incorporated ECP's are listed in this block for revision A of the specification)
③	④

Note. The effectivity is omitted because only one system is used.

Figure 18.3   System configuration chart.

Publication and distribution of the index may be specified and controlled by the customer, who may also indicate which SCN's shall be listed. A CCB directive could be used to specify customer requirements.

18.3 SYSTEM CONFIGURATION CHART

The system configuration chart is a summary record that identifies approved engineering change proposals (ECP's) with individual revisions to the system specification. This chart is issued by the company as a part of a specification revision. The chart is inserted between the title page and the first page of the specification. Refer to Figure 18.3 for a sample chart. The specification revision letter is placed in the first column, block 1. All ECP's incorporated in the revised specification after its initial release and current with the "A" issue are recorded in block 2. The next revision, "B," is placed in block 3 and ECP's incorporated in revision "B" are listed in block 4.

18.4 EQUIPMENT (END ITEM) CONFIGURATION CHART

The end item (CI) configuration chart (see Figure 18.4) is a summary record of approved ECP's with individual revisions to the equipment (CI or end item) specification. The chart is issued as directed by the customer and is inserted between the title and first page of the specification.

An end item configuration chart is prepared with each revision of Part I or Part II of the equipment specification. The chart contains the same data as the system configuration chart, except that a production effectivity column is added for recording the serial numbers of the equipments affected. The principal difference between the two charts, system and end item configuration, is that the system chart applies to changes to the system specification document only while the end item chart covers changes to the CI specification. As with the system configuration chart, the "ECP's" column includes all the ECP's that are (a) approved for the CI after the previous specification issue was released and (b) incorporated in the current issue.

18.5 SUMMARY

Engineering changes or revisions to specifications are implemented with SCN's and the documents described in the preceding sections. Normally, changes to specifications are made by attaching these records, with the exception of the specification identification index, to released documents. When changes are excessive, and a specification revision is authorized by the customer, the changes described in the SCN's are incorporated and a new document is issued and identified by adding the next letter of the

ABC Corporation Washington, D.C.	Code identification number 09876	End Item Configuration Chart	Date _____ Superseding (Issue date of last chart)

Specification number _____ Equipment nomenclature_____ CI number _____

Specification Issue (revision and date)	Engineering Change Proposals	Production Effectivity
A

4/15/71 | (Incorporated ECP's are listed in this block for revision A of the specification) | (Effectivity of specification revision on production item serial numbers) |
| B

7/22/71 | (Incorporated ECP's are listed in this block for revision B of the specification) | (Effectivity of specification revision on production item serial numbers) |

Figure 18.4 End item configuration chart.

alphabet after the specification identification number. The change log and configuration charts are updated if necessary and included in the revised document to provide a convenient historical reference of all the changes made to the specification for each revision letter.

A summary of the records used for changing or revising specifications follows:

Record Title	Function/Application	Responsible Party	Remarks
SCN	Definitive record for all major changes[2] to specifications; attached to specification	Originator of specification	Key change record
Specification change log	Cover sheet for SCN or a group of SCN's; attached to specification	Originator of specification	Interrelates SCN and ECP numbers, pages changed, and CI's affected.
Specification identification index	Summary of all specifications for a system; not attached to specification	Integrating or systems engineering contractor	Release is controlled by the customer
System configuration chart	Summary of ECP's approved against the system specification; attached to specification	Integrating or systems engineering contractor	Release is controlled by the customer
End item configuration chart	Summary of ECP's approved against a CI specification; attached to specification	Originator of specification	Release is controlled by the customer

[2] Minor editorial or typographical errors are not corrected until the next SCN is prepared for a technical change to the specification.

Chapter 19

ENGINEERING CHANGE AND CONFIGURATION RECORDS

Engineering change and configuration records are described in this chapter. Although there is a much larger variety of records in use than those included here, the records presented provide a good cross section of the documents currently employed by various companies. However, in presenting a representative series of format examples, there is always the danger of unavoidable redundancy between portions of the records illustrated. Therefore these examples should not be construed to present an optimum configuration accounting structure. Careful tailoring must be applied by the reader. The specific types of records selected for discussion are the engineering order status list; engineering change proposal list; configuration status and accounting index; engineering order data transmittal form; as-planned, as-designed, and as-modified lists; baseline configuration; parts lists; and a data traceability list. In addition this chapter covers: configuration difference list; file of approved EO's, deviations, and waivers; equipment log book; in-process test procedure index; waiver index; deviation index; and data index.

19.1 ENGINEERING ORDER STATUS LIST

An engineering order status list, as shown in Figure 19.1, is an index of all EO's prepared for the project and their status. (Note that ADCN, DCN, and so forth, can be substituted for an EO when these change control documents are used by the company.)

19.2 ENGINEERING CHANGE PROPOSAL STATUS LIST

The engineering change proposal (ECP) status list is similar to the EO status list. The ECP status list identifies all ECP's written by the company,

ABC Corporation Washington, D.C.

Code identification number 09876

Date ________
Superseding ______

Number | Rev.

Engineering Order Status Record

Project: ____________

Sheet of

EO Number	Title	Type[a]	Drawing Number	Processing Status					Change identification number / CI name
				Design	CMO	Change Planning	Customer Approval	Release Date	

[a] *Type.*

M—Mandatory
I—Improvement
R—Record
X—Expedite
DV—Deviation
WV—Waiver
SO—Stop order

Figure 19.1 Engineering order status record.

submitted for approval, and approved by the customer. It also includes associate contractor ECP's related to the company's ECP's.

19.3   CONFIGURATION IDENTIFICATION AND ACCOUNTING INDEX

The configuration identification and accounting index (report) lists all EO's applicable to an individual CI and the drawings and parts affected by the EO's listed. This report gives a summary of all key data related to CI packaged changes. For a large project, the report may be a computer print-out similar in format to that shown in Figure 19.2. In this case, computer inputs, prepared by the CMO, are converted to a complete set of 3 1/4-by-7 3/8-inch IBM computer cards similar to Figure 19.3. Each card is keyed to one EO and contains all applicable data in the form of punched holes. A card similar to that shown in Figure 19.4 may be used to authorize and instruct the computer facility to process and print out the index. A description of the contents of a sample index page follows.

The print-out for a flight equipment might be 100 pages on 11-by-13 1/2-inch paper. The order of listing is in increasing magnitude of the part or drawing number. When changes apply to the same number, the EO's are listed in order of increasing revision letter; that is, A first, B second, C third, and D fourth.

Note that abbreviations are used extensively because the IBM cards are limited to 80 characters or columns. Thus a compact descriptive system is required to record all significant data on one card.

Each page of the print-out is identified with the project name, company identification code number, and legends for the coding systems used. The date of print-out should be on at least the first page along with the computer program number used for processing the cards and providing the printout.

The list is updated biweekly or as required by the project manager or contract requirements. A data transmittal sheet is used for recording data as the EO's are released by data control and for providing the key-punch operator with a convenient input for typing or punching data onto the IBM cards. This data transmittal form is discussed in the next section.

The second type of configuration index is shown in Figure 19.5. This index identifies the status of all ECP's applied to an individual CI. In this case, all ECP numbers written against CI part number 123456-110 are listed on this index. Note that additional columns are included for the contract change notice, modification or retrofit work order number, spares affected, and new part number assignment.

The exact form of the configuration identification and accounting index can take many shapes, depending on project or customer requirements. Many

		Number	Rev.
(ABC) Corporation Washington, D.C.	Code identification number 09876		

Configuration Identification and Accounting Index—EO Status

Project: _____

Date _____
Superseding _____

Sheet ___ of ___

EO Number	Title	T^a y p e	Drawing Number^b	Part Number Affected	CI Name	Effectivity		Related Data Affected	Change Identi- fication Number	EO Release Date
						From	Through			

^{a} *Type.*

M—Mandatory
I—Improvement
R—Record
X—Expedite
DV—Deviation
WV—Waiver
SO—Stop order

^{b} An additional column for the next higher assembly could be added if desired.

Figure 19.2 Configuration identification and accounting index (EO status).

Note: Card shown is for illustration only. Punched holes do not correspond to letters shown.

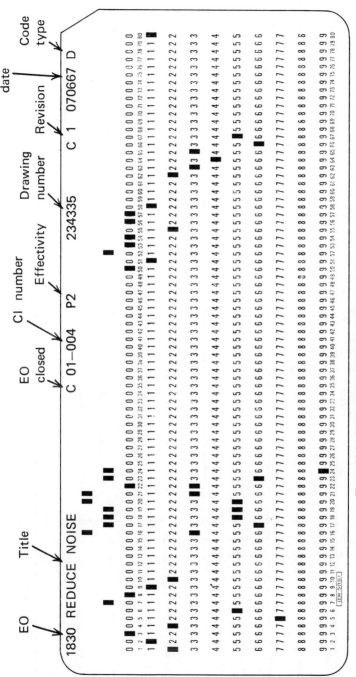

Figure 19.3 IBM card used for data input to configuration index.

Name	Department	Date	Job	Call	System
T. Samaras	640	21 June 71	CIN[a] status	246	PSA

Disk number ______

Punch ◯

Bin ______ BCD ______

Print ◯

Number records ______

Plot ◯

Number files ______

Number across ______

Estimated time ______ hrs

Meter time (hrs.)				Elapsed time

Special operating instructions

1. Primary sort—alphanumeric by part number
2. Secondary sort—alphanumeric by EO rev. letter

(Change date on pages)

(six-ply paper—6 copies)

Execute (X)

Keypunch ◯

Reproduce ◯

List ◯

Interpret ◯

Sort (X)

Number records ______

[a] Change identification number.

Figure 19.4 Instruction card to computer facility.

ABC Corporation Washington, D.C. | Code identification number 09876 | Number | Rev.

Configuration Identification and Accounting Index—ECP Status

Project: _____

Date _____
Superseding _____
CI number _____

Sheet ___ of ___

CI P/N 123456-110   CI name _____

ECP Number	Title	Seqª	Effectivity				ECP Submittal Date	ECP Approval Date	CCN Number	Mod/Retrofit Work Order Number	Spares Affected?	Change Incorporation Date	Part Number(s) Affected
			Retrofit		Production								
			First	Last	First	Last							

ª Sequence of incorporation

A. Before next systems test

B. Before mating with system at customer's facility

C. Launch site retrofit, etc.

Figure 19.5 Configuration identification and accounting index (ECP status).

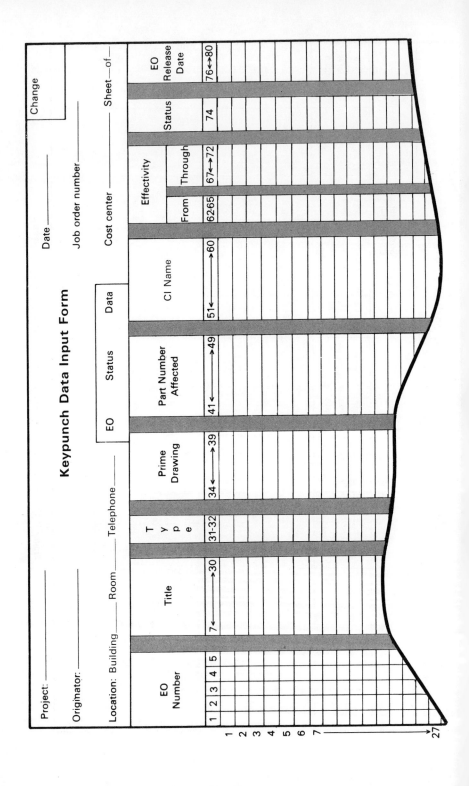

projects require separate status reports for spares, modifications, technical orders, and so forth.

19.4 CONFIGURATION IDENTIFICATION NUMBER DATA TRANSMITTAL FORM

The configuration identification number (CIN) data transmittal form is shown in Figure 19.6. This form is used to transmit data from engineering to the computer facility. The layout conforms to the IBM card requirements and consists of 80 columns and 27 rows. Each row represents a card entry. Since only 24 spaces are available for the title, the description words must be abbreviated because most EO titles will exceed 24 letters.

19.5 INTEGRATED RECORD SYSTEM FOR CONFIGURATION STATUS REPORTING

Chapter 3 discussed the need for a central "bank" of configuration records which grows with the development of the product. The most difficult task in maintaining a current representation of the product configuration is preserving the integrity of the indentured relationships and cross-relationships of the thousands of parts, assemblies, inspection and test data items, and so forth, that form today's typical product.

Figure 19.7 is an abstract representation of a system showing the logic that generates indentured data and parts lists. The subelement identifiers shown represent indenture levels and can be used to unambiguously address a specific item. One of the fundamental problems in configuration control is this:

Given that EC/DB is noninterchangeable modified, similarly affecting EC/D, but leaving modified EC/ interchangeable with all E/ and above, how do you identify A/ as containing modified EC/DB?

If A/ contains many thousands of parts, the classical task of changing all higher level prints for every part change is clearly unacceptable. The method of data management specifies that if EC/ is completely interchangeable in "form, fit, and function" into E/, no indication of change is apparent at E/ or above, and in many systems even EC/ bears no indication of the change.

If the change to EC/DB was "mandatory" there is an urgent need to know that all A/ and E/ contain the new EC/DB. This problem is solved by some form of serial number effectivity, part number, and dash number techniques. (See Chapters 11 and 12.) However this method breaks down completely on nonserialized parts, and is responsible for the plethora of serialization

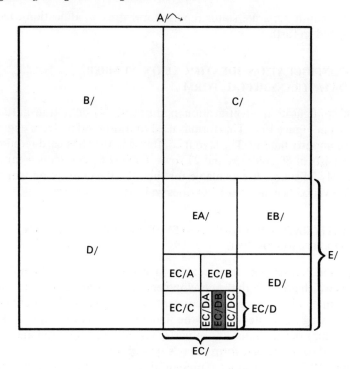

Figure 19.7   Product indenture model.

requirements and significant part number schemes on many projects. Because of the inevitable changes in the detail of the indenture record, it is best handled on electronic data processing equipment (EDP). The basic record must facilitate the baseline information inputs directed to it as the product's development moves through planning, design, fabrication, test and acceptance, and finally modification.

Given an indentured configuration record (either on file or in computer memory) that technically describes the approved design of a product completely and with currency, there remains the difficult task of verifying the actual as-built hardware configuration. (See Chapter 16.) The types and quantity of manufacturing inspection and test data are enormous. The techniques to be applied in the collection of this data are many. The following rather "gross" computerization approach is presented as one possible technique.

The as-built computer input is best created on a set of cards. Much of the data required on the card can be prepunched, the remainder being punched in on the spot by the inspector. Portable card punches are now

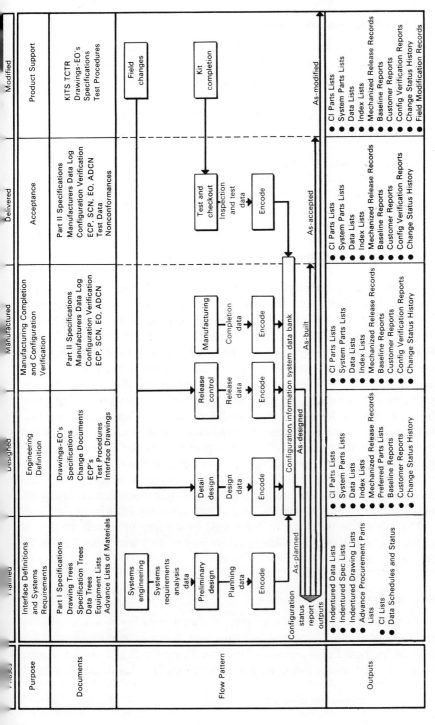

Figure 19.8 Integrated configuration record system.

available for this kind of work. The data needed for eventual hardware checking are part of those needed in the quality data gathering system, and can therefore be picked up as a bonus from a sophisticated quality data system.

Information needed for hardware configuration system reporting includes the following:

1. Part number.
2. Drawing number and revision.
3. Engineering indentured record address number (as-approved record).
4. Computer sequence number.
5. Item noun name.
6. Quantity.
7. Serial numbers (if applicable).

Each time an item identified on the "as-approved" indentured record is accepted in its final form, the acceptor creates a card with the information shown above. The part number (name and engineering record address) is already *prepunched* on the card selected; the computer sequence number is punched from the verification sheet the acceptor used; quantity, part revision letter,[1] and serial numbers are also entered at the time of quality acceptance. Depending upon the configuration stability of the product under manufacture, the card can go immediately to data processing or can be kept with the hardware until final acceptance. When a complete deck is believed to have been accumulated, comparison with the as-approved type can be quickly done and a status report printed out. Discrepancies can be worked out until complete agreement or disposition is achieved.

The known shipped configuration can be stored or transmitted in tape form as required by the contract. Future modifications can be easily and quickly inserted either from the field or at maintenance facilities.

The typical outputs of an integrated configuration record system, as previously discussed, are illustrated in Figure 19.8 and described in the following sections.

19.6   AS-PLANNED LIST

As-planned lists are used to enable advanced planning by manufacturing, production control, material control, and logistics departments. These lists are prepared on the basis of the specification tree and design requirements defined in the equipment specification. The drawings are initially

---

[1] Same as part drawing and revision letter.

identified in an indentured assembly drawing list. (See Chapter 3.) Detail drawings are added to the list as the task progresses. The list is released by engineering and authorizes the release of drawings that are specified on the list.

19.7 AS-DESIGNED LIST

When a drawing listed on the as-planned list is formally released, the drawing constitutes the as-designed configuration requirement. The purpose of this list is to provide a reference point for comparing the as-built configuration to the as-designed configuration. Discrepancies indicate incorrect drawings or manufacturing instructions or mistakes in the as-designed or as-built lists.

19.8 AS-BUILT LIST

An as-built list is a record of the configuration of a specific equipment serial number. The list identifies each major subassembly used in the equipment and the drawing and revision letter to which the subassembly was built. The lowest equipment level identified depends on contract requirements and could go to the piece part level. Additional data include part numbers, information indicating whether each of the subassemblies listed is interchangeable with the baseline configuration approved by the customer during FACI, and the identification of manufacturing and purchasing traceability documents. A major subassembly as-built format for an equipment is shown in Figure 19.9. Identification of each subassembly is made in the list.

The list is prepared by the CMO and validated by the quality assurance engineer. The list can be made from an inspection of the subassemblies as they are assembled into the equipment or from manufacturing records that identify the drawing numbers, revision letters, and serial numbers of every subassembly put into the equipment.

19.9 AS-MODIFIED LIST

An as-modified list identifies the changes made to a specific equipment in the field or at the test site. This report therefore provides an updated record of the equipment configuration as it proceeds through a series of field modifications or changes. The report includes the CI and serial numbers, modification drawing number, engineering change proposal number, date of incorporation, location of equipment modified, persons making the modification, and the results of the modification.

As-built Configuration Status Record

ABC Corporation, Washington, D.C. Code identification number 09876

Number ___ Rev. ___

Contract number ___

Project ___ CI part number ___ Date ___ Sheet ___ of ___

CI Name ___ CI number ___ CI S/N ___

Item or Sequence Number	Change Code	Assembly Level	Part Number	Description	Quantity per NHA	Next Higher Assembly Part Number	Acc. Rev. Ltr.[a]	Mfg. Rev. Ltr.[b]	Mfg. Ser. No.	Mfg. Lot. No.	Purchase Order Number	Manufacturing Shop Order Number

Preprinted format. Mechanized data lists may use computer-printed header and title information without preprinted guides

[a] Drawing revision letter acceptable to engineering for building item.
[b] Manufactured-to revision letter.

Figure 19.9 As-built configuration status record.

19.10   PARTS USAGE LISTS

Parts usage lists are records of the parts used to build the equipment. These lists are prepared by the configuration manager and designer and can be presented in two forms:

1. Parts usage by part number.
2. Parts usage by next assembly.

Parts usage by part number is a listing of equipment parts in the alphanumeric order of the parts. This is equivalent to the systems parts list described in Chapter 3. The order follows the alphanumeric numbers listed in the left-hand column marked "Part No."—letters first, numbers next.

The parts usage by next assembly is identical to the previous listing except that the parts are listed in order of their next higher assembly number.

19.11   DATA TRACEABILITY LIST

The data traceability list is used to provide data on parts, items, or materials bought from a vendor. This document lists all purchased and manufactured parts in part number sequence with a source code for each vendor and the lot or serial number. In addition, the document lists the CI number into which each part went and the CI's geographical location. Thus each installed part can be traced to its CI and to its location. Special dispositioned (for example, scrapped) items are also identified in the list.

19.12   CONFIGURATION DIFFERENCE LIST

The configuration difference list is a comparative listing of differences between a specific production equipment and the first production equipment (FACI) accepted by the customer. This document provides engineering and the customer with a quick reference for determining the differences among various production equipments and the authorizations for these differences. Difference data include the following:

1. Part numbers.
2. Drawing revision letters.
3. Quantity.
4. Traceability code.
5. Engineering orders, ADCN's, or DCN's released after FACI.

19.13   FILE OF APPROVED EO's, DEVIATIONS, AND WAIVERS

A file of customer approvals for EO's, deviations, and waivers is maintained by data control. This file includes teletypewriter exchange messages,

CI name _____ CI number _____ Serial number _____

Date	Accumulated Time Cycles	Test Type, Test Problems, Failures, Rework, Modifications, etc.	Applicable Documents
1/18/70	On 10:00 a.m.	A final acceptance test of the PTM was conducted according to acceptance test procedure	
	Off 11:45 a.m.	17168-TP-ATP. Input and output voltages were checked; regulation under no load and full load condition was checked; and ripple and noise checks were made.	
		The applied voltage/currents were 30.1 Vdc at 105 ma.	
1/18/70		R13 in the regulation circuit was replaced because of out-of-tolerance conditions (7% instead of 5%). In addition, transistor Q12 leads were reworked by J. Doe to eliminate possible shorts to R14 and C12.	Pfr 100 ISR 100 rework
1/19/70	On 1:00 p.m.	Power consumption was 3.19 watts, just under the maximum permissible of 3.2 watts. Retested	
	Off 2:30 p.m.	PTM	17168-TP
		The PTM met all performance requirements and is ready for delivery to customer.	
Total time	3 hr 15 min	Test engineer: Bill Jones 1/19/70	

contractual letters, contract change notices, and so on. The exact scope of this file is defined at the beginning of the project by the project and configuration managers. Although other project groups may be responsible for keeping these records, data control should have them available for convenient reference by the project manager and engineering staff.

19.14   EQUIPMENT LOG BOOK

An equipment log book is kept with each equipment, starting with its first assembly into one unit. This log book (see Figure 19.10) contains data on all the periods of operation, problems that developed during testing, failures that occurred, replaced components or subassemblies, persons conducting tests or rework, and dates that these tasks were performed. A brief narrative description of every operation is recorded by the test engineer, signed, and dated for each entry. The book usually has carbon sheets that are distributed to the engineer responsible for a company file and to the customer's representative. The original document accompanies the equipment after it is accepted by the customer. If the equipment is returned for rework, the log book is also returned. The extra copies of newly completed log sheets are removed and distributed before the equipment and log book are reshipped to the customer.

19.15   IN-PROCESS TEST PROCEDURES INDEX

An in-process test procedure index is an internal record of all in-process test and assembly procedures prepared for the project. This list identifies the document number, revision letter, title, and job or project number. The index is also dated so that superseding listings can be identified from old ones. Although not a formal part of configuration management, it provides a useful document for project control over revisions and status of these procedures. Of course, some projects may require formal configuration control, in which case the index is a responsibility of the configuration manager and is maintained and distributed as a regular control record.

19.16   WAIVER INDEX

A waiver index is a listing of all waivers written on the equipment delivered to the customer. A customer-approved waiver allows the company to deviate from contract or specification requirements. The information included in the list is the waiver number, the date, and its title. The waiver index may be eliminated by adding all waivers to the configuration identification and accounting index.

19.17 DEVIATION INDEX

A deviation index is a list of all deviations approved or rejected on equipments delivered to the customer. Deviations apply to engineering changes or to noncompliance of a specific equipment serial number to specifications before changes have been made. The deviation index may be eliminated by adding all deviations to the configuration identification and accounting index. The data recorded on the index are the same as those on the waiver index.

19.18 DATA INDEX

A data index is a tabulation of all engineering drawings, documents, associated parts lists, and specifications. It provides a convenient source for determining the documentation that is required for the project and its status. The order of listing is as follows:

1. Drawings.
2. Parts lists.
3. Specifications.
4. Standards.
5. Publications (reports, manuals, test procedures, and so forth).
6. Documents referenced on drawings and parts lists.

The index should include the manufacturer's code identification number (if different from the company's), drawing size, document number, sheet number, revision letter, and title or description. When a separate drawing list is issued periodically by engineering, project drawings can be omitted from the data index.

19.19 CONFIGURATION BASELINE INDEX

The configuration baseline index is a record of all approved engineering documents that constitute the baseline configuration of the CI. The index may include all drawings, specifications, parts lists, and test procedures used to design, build, and test the CI. Documents are usually listed in alpha-numeric order.

A preliminary baseline index may be prepared by the configuration manager for the critical design review and the final version can be released at FACI. The revision letter of each approved document listed must be recorded on the index. The configuration difference list can then be released periodically as the project progresses to identify document changes with regard to the original baseline index.

The reports and indexes described in this chapter contain all the information given in the baseline index. Therefore, the configuration manager or customer may not require the preparation of a baseline index or report for the project.

PART IV CONFIGURATION MANAGEMENT SUMMARY

- ● THE PRODUCT DEVELOPMENT CYCLE
- ● OPERATIONAL COMPUTER SOFTWARE CONTROLS
- ● THE CONTEMPORARY SCENE

Chapter 20

THE PRODUCT
DEVELOPMENT CYCLE

This chapter is intended to focus configuration management, as discussed thus far, against a backdrop of product development activities taking place throughout the life cycle of a typical aerospace product or system. In order to introduce the subject of product cycle activities, we have repeated in Figure 20.1 the illustrated baseline management concept presented in Chapter 1. It is important to note that the establishment of baselines must remain flexible and in consonance with the nature, scope, complexity, and stage of the life cycle of the item; and therefore baselines are an integral consideration to any discussion of the flow of product activities. Consequently the following subject matter deals with each of the major phases and baselines presented in Figure 20.1:

1. Conceptual phase.
2. Definition phase.
3. Acquisition phase.
   a. Preliminary design.
   b. Detail design.
   c. Qualification/Prototype fabrication and test.
   d. Production fabrication.

Although we have tried throughout the previous chapters not to become entrapped with overspecialized government or aerospace management situations which would not be in context with the *fundamentals approach* of this book, this chapter and the others in Part IV will tend to explore a number of specifics relating to the more complex aerospace systems and government requirements.

20.1 ESTABLISHMENT OF "SYSTEM" BASELINES

The process of baseline establishment and control, as presented in Chapter 1, for example, may appear quite straightforward when related

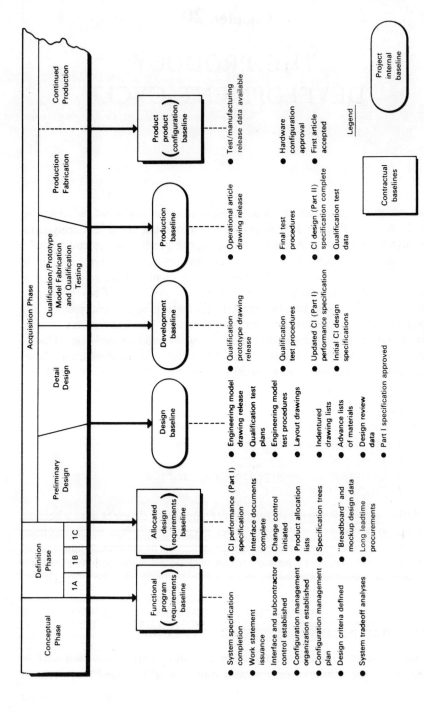

Figure 20.1 Product development cycle.

to an individual CI; however, the establishment of the baseline for an aerospace system is considerably more complex. The baseline must be established for an interweaving of products (CI's) that comprise an integrated spacecraft, air vehicle, or ground system, with all its attendant support requirements.

It is naive to suggest that an entire system can suddenly one day reach a particular baseline. Since the baseline is not a design freeze by any means, it can be, and usually is, established in an incremental manner. That is, each CI of the system configuration is separately audited against its particular optimum point of the life cycle of that particular element. After this, the audited item becomes subject to the formal configuration controls incumbent upon a baselined item.

Since many of the elements comprising a given system are supplied by associate contractors, subcontractors, vendors, and suppliers, and many items are government furnished, they must all be separately baselined and controlled. And so each baseline for the system becomes a set of discrete steps leading to a final audit which encompasses the baselined item and all documented changes to it.

In addition, consideration must be given to each item's complexity and its position in the work breakdown structure in order to adapt the baseline method to each specific item. In some instances simple control of an item's specification can satisfy all requirements of configuration management. In other cases, a full and systematic configuration management program must be employed. The time period, physical location, and complexity of item are all critical factors in determining the time and method of baseline derivation.

20.2 CONCEPTUAL PHASE

The conceptual phase of a project is directed towards demonstrating the *need* to undertake the development of a specific system or product. The government has formulated detailed, specific guidelines and requirements for the conduct of the conceptual phase. Therefore the following discussion is presented in the framework of government activities and terminology. Within the limits of the commercial environment, these same principles may also be applied to non-government projects to improve the efficiency and quality of new developments.

Frequently the initial development of a product has no clear starting point. The ideas generated in answer to a government need may evolve from senior staff teams responsible for planning future projects, high-level government conferences on the nation's goals, or significant advances in technical capability. They may also be brought into focus by industry

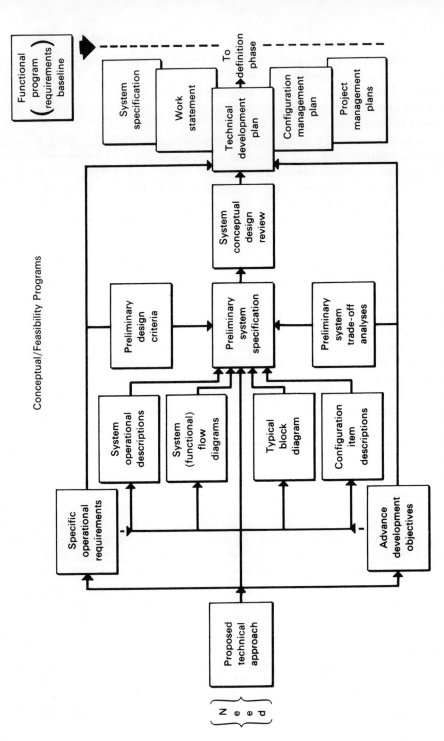

Figure 20.2 Product cycle—conceptual phase.

(for example, in the form of an unsolicited proposal outlining a new idea for a product). These ideas ultimately develop to meet the requirements of national objectives and technological capabilities, thereby precipitating a decision by the government to plan investment of resources: manpower, money, and facilities. In the final stage, research, exploratory efforts, or advanced developments lead to building experimental products to solve a recognized government need.

Conceptual phase activities, which consider future government requirements (5 to 10 years hence), and exploratory development, which shows promise of maturing and supporting the requirements of the time period, are examined exhaustively in feasibility studies. These studies verify the validity and practicability of the technical concepts, explore weaknesses and unknowns, and establish bases on which judgments can be made about the future of the effort. Therefore a sound technological base or hurdle for a proposed product must be assured and surmounted before the government is convinced of the project's worth.

In summary, the precipitating factor in the decision to begin development of a new product is usually one or more technical or scientific ideas. With these established, direct consideration must be given to such factors as the technical capabilities of other countries and the political situation; then a determination of what is desired, tempered with an evaluation of the current state-of-the-art, a forecast of technological feasibility, and consideration of economic factors.

Given a successful "conceptual" venture, the objective of all studies during this period may be either a government or a government-contractor *effort*. The objectives in both cases are the following:

1. To define product system requirements.
2. To determine feasibility and cost effectiveness.
3. To conduct trade-off analysis of cost and performance.
4. To prepare reports describing the theoretical, analytical, and technical aspects of the project.
5. To prepare cost estimates and schedule to develop and build the product.
6. To establish the first baseline: functional (program requirements) baseline.

In examining the block diagram of typical conceptual phase actions, shown in Figure 20.2, some parallels to commercial industry may be observed. For example, the long-range marketing plan of a company would influence its decision to proceed with the development of a particular product in much the same way that the government relies on its advance development objectives and future (5 to 10 years) program requirements. In private industry, in addition, the specific operational requirements may be considered to be the expression of the gross system or product requirements submitted by the research department; these requirements would serve as a departure point for various studies, investigations of alternate approaches, various physical, functional and cost optimizations, and so on, leading to a definitive technical agreement (system specification) for the product to be marketed. In summary, the objectives of the conceptual phase are axiomatic to good business: the market potential, cost effectiveness, producibility, and technological base for the planned product must be assured before engineering and production commitments are made toward its commercial development.

Direct configuration management actions are minimal during this period and are primarily directed towards the data control aspects of the system documentation as it becomes available.

20.3 THE DEFINITION PHASE

In the government-contractor efforts, the next phase of the product development is the definition phase, consisting of three distinct subphases. Phase 1A is a government effort which results in the selection of contractors to whom formal definition contracts (phase 1B) are awarded.[1]

It is in phase 1B of the definition effort that configuration management first becomes operative. Phase 1B is a government-funded contractor effort applicable to large development projects whose estimated research, development, test, and evaluation cost exceeds $25 million. In it, the products (black boxes) of the program (configuration items) are technically defined by systems analyses, performance specifications, specification trees, interface controls, major subcontractor agreements, breadboard and mockup models, engineering tests, and preliminary qualification test plans.

Thus, this phase is the beginning of so-called "configuration identification": the documentation of performance specification and interface requirements that will pursue a configuration item throughout its design, development, test, production, and service life.

Since configuration management applies to documentation as well as to hardware, the paperwork during this phase is also closely controlled. The first level of control is the system specification provided by the phase 1B work statement, which defines the tasks to be done and the products to be delivered to the buyer. Expansion and refinement of the system specification is part of the contractor's job during phase 1B and changes are common. The system specification changes in either of two ways. If the change is

---

[1] DOD Directive 3200.9, "Initiation of Engineering and Operational Systems Development."

buyer-originated, a directive is issued, accompanied by a specification change notice (SCN). If the contractor wants a change that affects the system specification, he submits a formal system requirements engineering change proposal (ECP) to the buyer. If approved, this leads to an SCN or a complete revision to the system specification. Other documentation produced and controlled during phase 1B includes the following:

● Realistic cost and schedule estimates and identification of high-risk areas.
● Firm systems operating, environmental, and performance criteria, supported by system optimization analyses.
● Definition of interfaces and corresponding responsibilities.
● Evaluation of time-cost-performance trade-offs (see Glossary).
● Subsystem performance requirements.
● A program breakdown structure (PBS),[2] or work order structure, which serves as the basis for the contractor's detailed cost estimate for each CI designated by the contractor.
● Validation of the technical approaches to determine a firm fixed-price or fully structured incentive contract for the acquisition phase. (See Chapter 22.)

Phase 1C is the critical period during which competitive proposals for the development effort (acquisition phase) are evaluated by the government. This is the time contractors wait with crossed fingers, while the customer decides to:

1. Accept one or more proposals.
2. Abort the entire effort.
3. Request redefinition from one or more contractors.
4. Transfuse the better parts of two or more contractor proposals into a complete or limited redefinition effort per item 3.

The Government's Definition Phase in Retrospect. The government definition phase concept evolved from a transitional history of Air Force life-cycle management concepts as illustrated by Figure 20.3.

In the past some systems were almost 10 years in design, development, test, and production phases before entering Air Force operational inventory using the "fly-before-you-buy" systems concept (Figure 20.3a). As systems became more complicated, these separate phases began to expand to even longer leadtimes, causing many to be obsolete either before they entered the inventory or shortly thereafter. When the cold war began to warm, it

---

[2] Same as work breakdown structure.

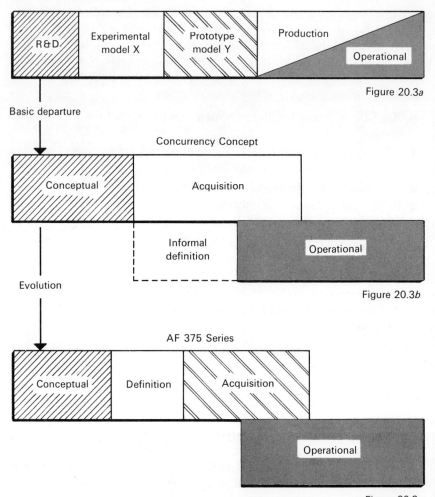

Figure 20.3a

Basic departure

Concurrency Concept

Evolution

Figure 20.3b

AF 375 Series

Figure 20.3c

Figure 20.3   System life cycles.

forced a speed-up in fielding weapons to complex high readiness capabilities, which resulted in the concurrency concept (Figure 20.3b), in which production is started on a system product concurrent with development, test, and construction of others. This concept called attention to the need for efficient and effective integration of simultaneous activities. The Minuteman ballistics missile is a prime example of the concurrency system concept. Today, concurrency is out; it is too expensive. The question even arises whether it is the fastest way to build a system.

To overcome objectionable features of the earlier concepts, the Defense Department (DOD) provided firm guidance that formalized the definition phase as a period during which government and competitive companies in industry analyze the requirements and refine plans for acquiring the system before committing engineering development dollars (Figure 20.3c); that is, the main feature of the definitive concept is a competitive program definition phase that follows firm establishment of the system but before any acquisition activity. Thus, introduction of the definition phase into the concurrency concept optimizes the best features of the old "fly-before-you-buy" and the pressured concurrency concept.

Contractor Definition Phase Activities. Whether the contractor is entering a government-funded phase 1B effort as described above or is advancing from his own conceptual phase activities, the significant project effort during the definition phase is to reevaluate the conceived design approach and establish the optimum subsystem design and its allocated configuration item performance requirements. As illustrated by Figure 20.4, activity during this period is high. Systems engineering analyses are at their peak; operational requirements are thoroughly considered; alternate design approaches are evaluated; engineering model criteria are formalized; selected breadboard and mockup development is initiated; all culminating in a formal set of system, subsystem, and configuration item technical performance and interface specifications. Thus this phase provides the transition to a unified, integrated project team; and the functional organizations are addressing themselves to the problems of project performance with respect to staffing, planning, familiarization, and the commitment to engineering design release.

This unification and integration of the project's activities are brought together through the management and technical plans developed during the conceptual phase and implemented through a well-defined work order structure and project master schedule. At this point in the product cycle, the configuration management organization is chartered and staffed. An initial task of the CMO is to "officialize" the functional baseline documentation of the conceptual phase (system specification, technical development plan, and work statement) into an authoritative and controlled technical description of the product.

A major portion of this early period of the definition phase will of necessity be occupied with performing analyses and conducting studies to determine any secondary requirements to meet the system specification and thereby

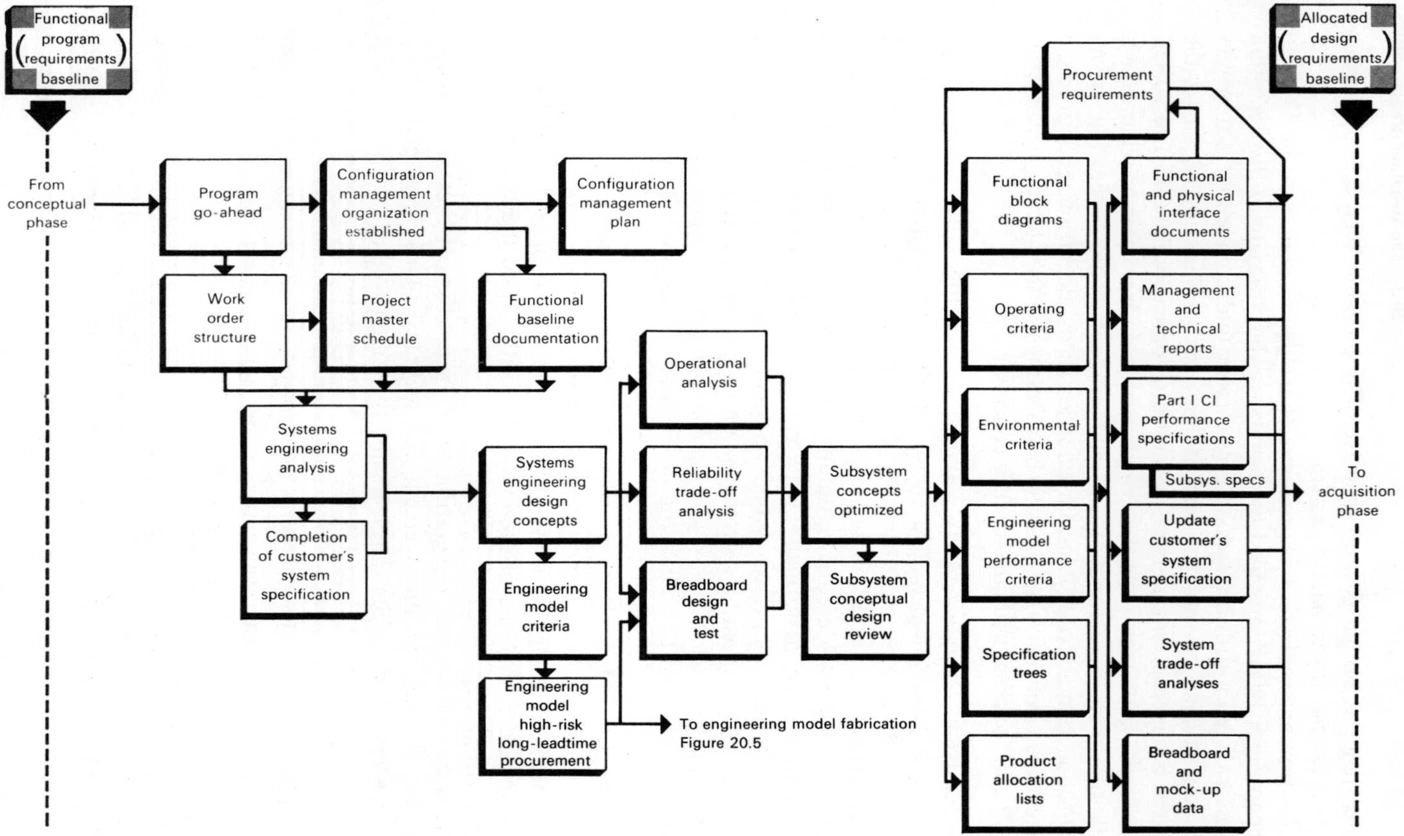

Figure 20.4 Product cycle—project definition phase.

expand and refine it. This objective is logically developed by taking the system specification and establishing the kind of analysis needed to determine the performance parameters required for the various subsystems and their configuration items and then ensuring that these subsystems are assigned their performance parameters through coordination between conceptual designers and design analysts. The importance of the systems engineering role in the optimization and analysis of design interface controls and firm systems design concepts cannot be overemphasized. Systems engineering must decide what the problems are, fit the parts of the solutions together, and thereupon commit and draw upon the skills of the hardware designer. Simon Ramo described the systems engineering role very well:

"Systems engineering is an old and very present phase of every engineering task. The job of integrating the whole, as distinct from the invention and design of its parts, the creation and analysis of the overall answer to a problem, the breaking down of the total into a set of harmonious specified parts, the assurance of compatibility and consistency in the ensemble, and the relating of that ensemble to the outside world that has originated the need and that will employ the final result—these considerations are present in varying degrees in every single piece or group of equipment, from a chair to a transcontinental railroad. Always some fraction of every engineering team has in effect been devoted to *this* systems engineering." [3]

The systems engineering tasks in the definition phase are a continuation of the preliminary systems engineering activities of the conceptual phase. In fact, the forming of the work order structure and work authority hierarchy of the definition phase was developed as an output of the systems engineering process of the contractor's own conceptual phase. In the case of the contractor entering a formal government phase 1B definition phase contract, he will expand a work breakdown structure provided by the government's phase 1B work statement in accordance with MIL-STD-881. [4] This expansion is also accomplished by the application of systems engineering. Therefore systems engineering, through its iterative process, contributes directly to the construction of product and task elements into a project work order structure.

The definition of systems engineering design concepts and product structure enables the entry into the basic packaging (layouts, sizings, and mountings) and circuit design for the engineering model. The engineering

[3] Simon Ramo, Vice Chairman of the Board, TRW Inc., "The New Emphasis on Systems Engineering," *Aeronautical Engineering Review*, April 1957.
[4] MIL-STD-881, "Work Breakdown Structures for Defense Material Items," November 1, 1968.

design approach is proved feasible by initial development breadboard testing. Data release at this point will include, as a minimum, the following:

1. Preliminary schematics sufficiently detailed to enable sizing estimates to be completed.
2. Identification of engineering model high risk and long-leadtime items, including part type and count.
3. Engineering model functional test levels and plans.
4. Layout constraints.
5. Engineering model performance criteria.

In parallel with the breadboard fabrication and test activities and the derivation of engineering model criteria, operational requirements are examined and the sensitivities of reliability trade-off analyses are reviewed. It is vitally important that the subsystem and CI designers understand the operational requirements and how they apply in practice to the "static" system requirements. An understanding of the operational requirements in detail can result in putting the design complexity where it is most appropriate for the system. It is also necessary to determine the reliability sensitivity of the proposed system/subsystem components. At this time, nominally simple system changes often result in extreme simplification of the subsystem design. Through reliability trade-off analyses, critical areas in design, manufacturing, and procurement are identified. It is imperative that potential problem areas be identified as soon as possible at this point in the design process to maximize the time available to resolve such problems.

Finally the system, subsystem, and CI performance concept is submitted to a second conceptual design review. The satisfactory completion of this design review releases the following documentation (Figure 20.4) which forms the allocated baseline:

1. Updated system specification.
2. Systems engineering functional block diagram.
3. Operating and environmental criteria.
4. Performance criteria for engineering model.
5. Specification trees.
6. Product allocation lists.
7. Formal interface documents.
8. Engineering model indentured drawing list.

Because the functional requirements of the subsystem and CI's are frequently changed during the course of this phase and because of the pressure of rapid changes on the schedule, the timely documentation of functional and performance requirements is important. The CMO responsibilities are therefore directed at implementing a clear and flexible engineering model documentation program and a responsive systems engineering data control capability, in conjunction with a formal control system for the system specification, Part I CI specifications, major subcontractor specifications and work statements, and interface control documentation. Engineering model identification requirements, the project specification tree, and product allocation list are also defined and implemented by the CMO before completion of the definition phase. At this point, we are ready to commit the engineering model design.

20.4   ACQUISITION PHASE

The acquisition phase is divided into five secondary phases: preliminary design, detail design, qualification/prototype fabrication and test, production fabrication, and operational. Each of these is discussed in the following sections of this chapter.

Preliminary Design Phase

The initiation of detailed engineering model design and test activities (Figure 20.5) occurs with the release of engineering model work authorizations and major subcontract work statements. There is considerable concurrency taking place between the fabrication and testing of breadboard, engineering model detail drawing preparation, and activities associated with the qualification model, prototype, and operational models during this period. For example, the support equipment concepts for *all* model configurations are usually determined at this point. This obviously is also true of the Part I CI performance specification which must represent the performance criteria for the qualification, prototype, *and* operational articles. Factory test equipment as well as "field" test equipment requirements are established concurrently for engineering model test support and for the delivered CI operational test program. Preliminary prototype, qualification, and operational layout drawings are initiated and form the heart of the interface control drawing structure.

The configuration management office assumes a strong role in the definition of the project's authorized list of interface control documents and their control. The configuration manager is a member of the interface control working group and acts as the control point for all interface change actions to mockup data and interface control documentation.

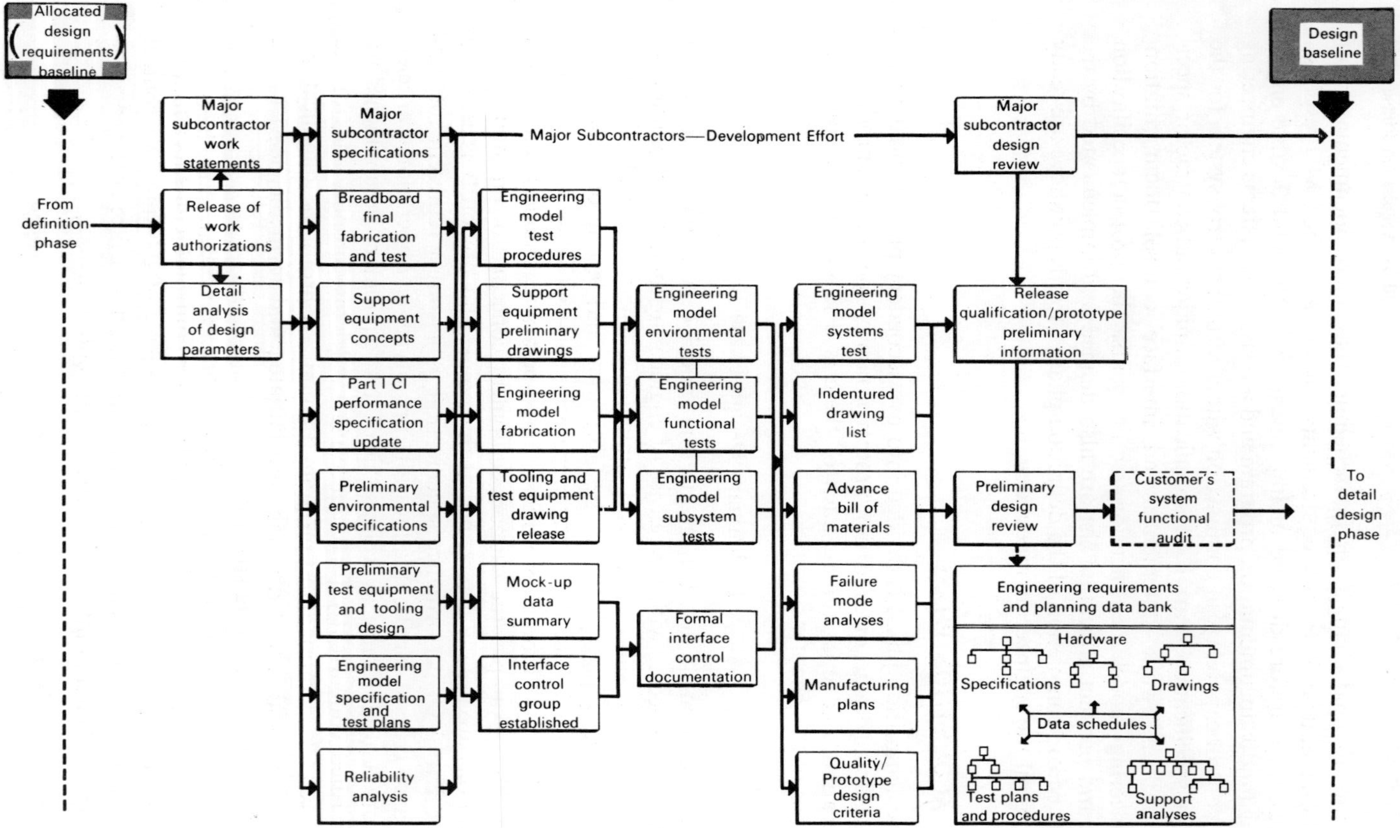

Figure 20.5. Product cycle acquisition phase (preliminary design).

The preliminary design phase concludes with the release of a design baseline consisting of the following approved documentation:

1. Qualification, prototype, operational indentured drawing lists.
2. Advance bill of materials for qual/proto development.
3. Engineering model test data.
4. Qual/Proto design criteria.
5. Final Part I CI performance specification.

At this point, the contractor's preliminary design review is held to authorize the initiation of detail qual/proto/operational design and development. Under government (Air Force) contracts, an optional customer-directed system functional audit (SFA) may also be required. The commitment to the qual/proto/operational development activity is formalized by entry of design planning information into the configuration data bank: that is, specification and associated data, indenture information, hardware allocation planning, drawing structures, test procedure trees, and supporting analysis documentation.

Detail Design Phase

A significant element of the activities associated with the detail design phase (Figure 20.6) is the final determination of those items to be manufactured by the contractor and those to be subcontracted. Based on these decisions, the qual/proto/operational detail drawing development is initiated. Part I CI performance specifications are once again updated. Final layout drawings are approved and released. Procurement documentation and planning are refined. Engineering development concentrates on the preparation of assembly and subassembly drawings, mechanical detail drawings and master artwork, schematics and wiring diagrams, and the test procedure program. Part II CI product specifications appear here in their initial form for the first time.

This phase concludes with the release of qual/proto design for fabrication and test. As defined in Chapter 14, an "internal" development design review is held for the purpose of formalizing the qualification and prototype configuration and planning committed to the contractual qualification test requirements; thus we enter the qual/proto fabrication and test phase.[5]

During this phase, CMO change actions are directed at the qualification and prototype drawing configuration, the qualification test program, and overall procurement activities. The CMO constantly establishes indices and schedules for specification, drawing, and test procedure releases in order

---

[5] Note that for some projects, the CDR may be held after engineering model evaluation and before qualification or prototype model fabrication is begun.

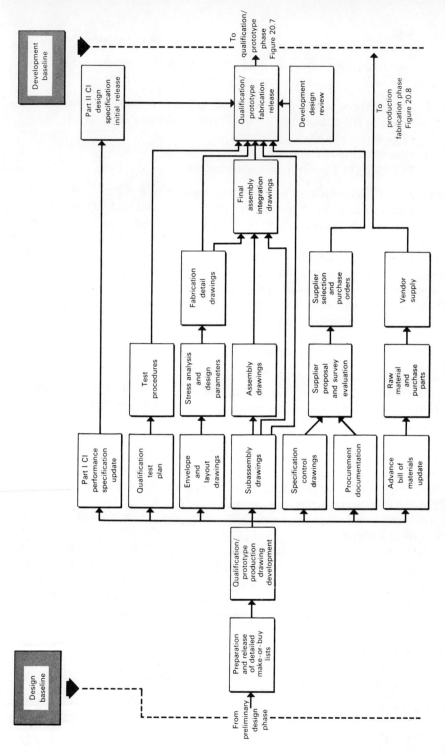

Figure 20.6 Product cycle—acquisition phase (detail design).

to assure effective configuration identification. The interface control documentation is updated as required. As-built fabrication and test information is being accumulated and verified.

Qualification/Prototype Fabrication and Test Phase

The CMO, during the activities illustrated in Figure 20.7, continues to ensure the accumulation and management of as-built historical product data generated during the fabrication, assembly, inspection, and test of the qual/proto models. A large number of mandatory changes against the design and test program are reviewed and controlled by the CMO during this period.

The accumulation of documents depicting the configuration, manufacturing, and inspection history of the qual/proto articles, their assemblies and parts, is accomplished by "manufacturing data package" control. (See Chapter 16.) The manufacturing and test data record is identified to the material to which it pertains and accounts for all manufacturing and inspection operations performed on the material, including any non-conformances, failures, parts replacements, and modifications. The controlled application of serialization requirements is essential at this time.

The manufacturing, assembly, inspection, and test surveillance operations which reflect the manufacturing, configuration, serialization, or quality history are recorded on the "in-line" manufacturing shop orders (manufacturing orders) and parts lists, copies of which must be included in the as-built configuration record and data bank as prescribed by the CMO.

The manufacturing data package must accompany the material with which it is associated at all times, including rework and material review disposition activities. When the last inspection operation for a part or assembly is completed, the quality inspector verifies that copies of all applicable documents are accounted for, validates the data package, and forwards it to the as-built record center or mechanized data bank.

The Part II CI design specification is completed at this time along with a final update of the prototype configuration drawings into the operational article requirements. The contractor preproduction design review is held. A customer review (critical design review or functional configuration audit—DOD) may also be held, as described in Chapter 14, before releasing the production configuration.[5]

Production Fabrication Phase

Changes from this point forward assume added importance, since full production and test resources are involved. Consequently, *all* changes are

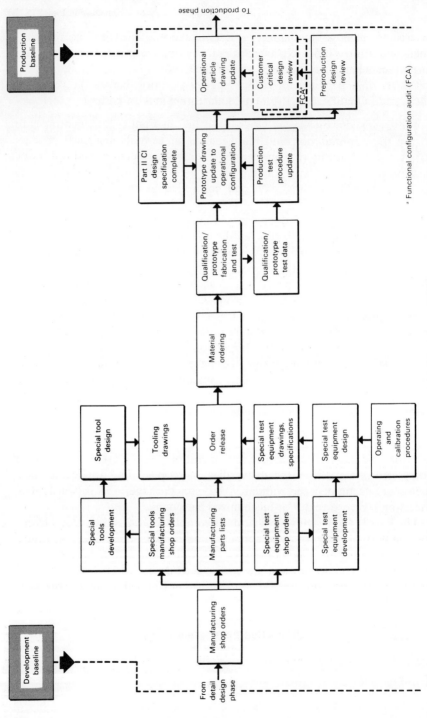

Figure 20.7 Product cycle—Acquisition phase (qualification prototype fabrication and test).

processed by engineering change requests to the CMO, except for record changes and in those cases of proven emergency where expedited engineering change documents must be issued. The CMO will require notification and justification of a pending emergency change situation before the release of the expedited engineering change document (EO, ADCN, and so forth). The change document must receive the approval and formal release of the CMO as soon as possible after the commitment of the expedited change to manufacturing.

The detail activities during this phase (Figure 20.8) are the same as described in the qualification/prototype fabrication and test phase (Figure 20.7). In fact, there is a considerable and natural overlap of production, prototype, and qualification CI fabrication activities occurring during these phases. Many production CI's which have not experienced major configuration differences with respect to the qualification/prototype CI may be fabricated during the qual/proto configuration fabrication period. This is also true of long-leadtime items which require early procurement and fabrication action.

The configuration of the first acceptable production article establishes the most meaningful of all baselines: the "customer" or "product baseline." It is the baseline recognized by the buyer, by means of FACI or PCA procedures, as the point of departure for all configuration changes to both specifications and drawings throughout the acquisition and operational phases of the project. All the disciplines of configuration management are now in effect for the life of the contract, including fully detailed configuration reporting and customer surveillance over all changes.

Operational Phase

This phase begins when the first system is accepted by the buyer. The operational phase includes such commonly used terms as the "acceptance phase," the "deployment phase," and the "product utilization phase." All such phases after FACI/PCA continue to be related back to the customer product baseline. During this phase, CI's are being concurrently produced and shipped according to the customer's allocation plan. In missile and spacecraft systems, an installation and checkout procedure is generally involved, followed by turnover to the procuring agency and site activation. In many cases of military procurement, the CI is subjected to operational testing in an integrated system.

The operational phase extends over the useful life of the CI. In the case of spacecraft or satellites, the phase may end at launch; in other projects there will be a continuing responsibility for tracking, telemetry, and data reduction. In the case of weapons systems or equipments, the operational

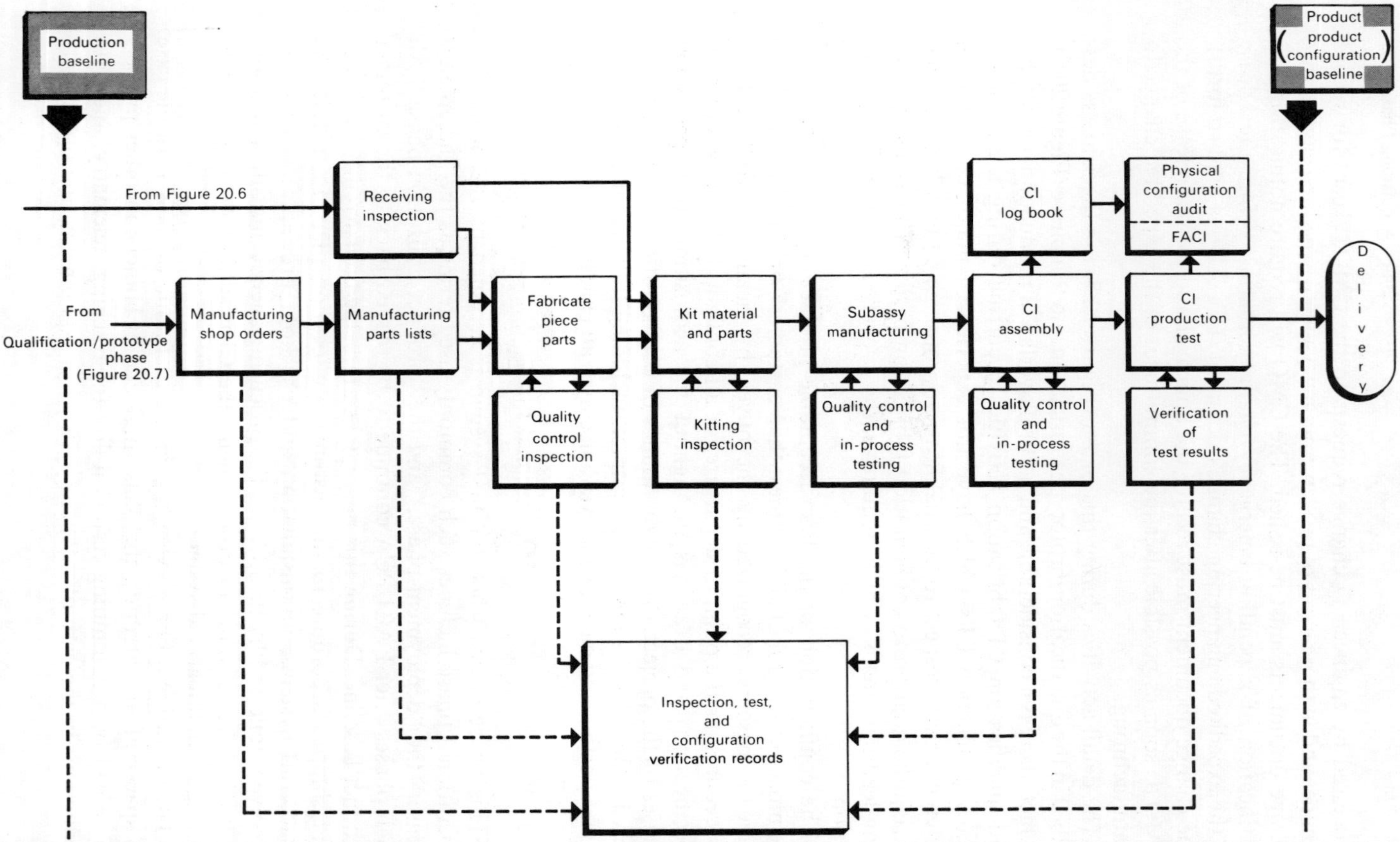
Production baseline
From Figure 20.6
Receiving inspection
From Qualification/prototype phase (Figure 20.7)
Manufacturing shop orders
Manufacturing parts lists
Fabricate piece parts
Quality control inspection
Kit material and parts
Kitting inspection
Subassy manufacturing
Quality control and in-process testing
CI log book
CI assembly
Quality control and in-process testing
Physical configuration audit
FACI
CI production test
Verification of test results
Product (product configuration) baseline
Delivery
Inspection, test, and configuration verification records

phase lasts until the items are phased out of the military inventory. These phases involve no new baselines, since full configuration management has been instituted at the product baseline. The operational phase is characterized by gradually decreasing contractor participation in configuration management as the product configuration is optimized and stabilized by retrofit ECP action.

Note that the operational phase has been included here as a part of the acquisition phase because the two phases often overlap. However, the operational phase is normally a separate and fourth phase of a product's life cycle.

Chapter 21

CONFIGURATION MANAGEMENT OF OPERATIONAL COMPUTER PROGRAMS

A cursory study of some of the software systems developed before 1968 will reveal that the management techniques employed were not well planned or integrated during this period. In recent years there is evidence of a change toward a more formalized management concept, especially by the more progressive practitioners. In particular, managers in the computer-software community who experienced schedule and cost overruns are now moving toward the methods and procedures familiar to hardware systems development.

The problems experienced by the early developers of software systems were due to two primary elements:

1. Failure to use the planning and control techniques available.
2. The unique and new problems presented by software management.

In the past 3 to 4 years many developers started taking advantages of some of the management techniques depicted in various government documents, such as PERT[1] scheduling and formalized specifications programs. So far, however, not many of these concepts have been employed in depth nor stressed during the appropriate project life-cycle phases. Although these tools were originated and have been progressively improved for hardware project application, they are equally applicable in principle to software.

---

[1] PERT: Program Evaluation and Review Technique. The PERT system is a tool for determining at any point in the life of a project what the status of each key task is and where the problem areas are. PERT was developed to provide effective control of complex projects when the traditional methods were proving ineffective. PERT is used in commercial and government work.

Blended systematically with the recognized need to concentrate on the inherently new and special aspects of software management, formal and consistent implementation of systems engineering, provided by the disciplines of configuration management, and the like, will have significant impact on the final outcome of software projects.

Basically there are two distinct differences in management of hardware as compared to software projects that alter the relative importance of their particular life-cycle phases. This is especially true for the software planning, control, and integration aspects. These differences are the following:

1. Software is a logical arrangement of symbolic data and information using many varied techniques, theories, and methods, for the most part conceived and stored in the minds of men.

2. Hardware is basically comprised of physical objects, piece parts, components, and materials assembled on common or associated chassis or platforms by using, for the most part, documented methods and processes, many of which have long been accepted as universal standards.

These basic differences can be further expanded for the following reasons:

1. Hardware is joined together by mechanical and electrical interfaces which permit some tolerances in form, fit, and function.

2. Software is composed of symbolic substance and therefore admits no tolerances comparable to those of hardware; the part-to-part relationship of software is logical in nature and absolute, either true or false.

3. These precise software traits and unique requirements of software systems themselves compound the development of valid testing programs into tremendous undertakings. These frequently can be of the same magnitude as the development of the basic system design. Traditional testing techniques, such as sampling, visual inspection, and measurement of properties, that worked so well on hardware are practically useless on software. Validation of software test plans and procedures can be a time-consuming endeavor requiring excessive labor, machine time, and rework during the checkout and debugging phases, especially where effective techniques and controls were not employed to assure orderly and cost effective progress through the definition phase.

4. Hardware systems frequently contain many off-the-shelf and standard parts, materials, and processes whereas software projects can easily contain no comparable readily available or standardized elements.

5. Finally, documentation of hardware has become greatly standardized whereas software is about where hardware documentation was 10 to 15 years ago. How much documentation is enough and in what form and content are questions for which answers are still in formative and evolutionary stages.

Looking at this from the project and configuration management standpoints, then, major hardware problems fall into the integration of production and logistics with the design and development phases, especially where such things as reproduction of like models, maintenance, and spares support requirements are factors important to the customer. Software project management's greatest concern is in the integration of system test with the design and development phases of the system. Although this shift of concern and interest may seem slight on the surface when comparing software versus hardware management, the ultimate difference in schedules and cost can be significant if adequate controls are not provided for the system test phase early in any software project life cycle.

The guidelines presented here describe primarily configuration management, product integrity requirements, and procedures for the control of computer software programs. Particular emphasis is applied to the following:

1. Structure and content of the technical computer program documentation.

2. Description of a numbering system for the identification of computer programs and documentation.

3. Control, documentation, and dissemination of engineering changes to the technical computer program documentation and computer program end products.

4. Technical reviews and inspections during the development cycle.

5. Quality assurance test validation, acceptance, and verification of computer program end products.

21.1 COMPUTER PROGRAM CONFIGURATION ITEMS (CPCI)

The key concept of software development controls is the timely generation and control, as information becomes available, of increasingly more detailed software development documentation describing the operational computer program, its basic elements, and end products (computer program configuration items, or CPCI's).

The CPCI is the essential item of management concern and must therefore be thoroughly identified and specified. For management purposes, any given CPCI may be divided into further subdivisions (modules) and designated as computer program components (CPC). The summation of all individual CPCI's (and CPC's) is in fact the operational computer program.

For purposes of configuration management, the CPCI is defined as a punched deck of cards, magnetic tapes, or other physical media containing a sequence of instructions and data in a form suitable for insertion into a digital computer. Thus, two separate and distinct aspects of the CPCI configuration may be inferred, having different implications for management:

1. The physical form, dimensions, and materials of the tape or card deck media.
2. The actual sequence and content of the instructions and related data.

Since the tape or card deck medium must be compatible with input-output equipment attached to the computer and represents the physical object which is delivered to and accepted by the user, this aspect of the CPCI configuration cannot be altogether ignored during the development process. However, the requirements are relatively routine and lend themselves to detailed standardization without regard to the particular content of a given CPCI. In general, it is the second aspect of computer program configuration—namely, the actual content—with which the provisions contained herein are primarily concerned.

Selection of Computer Program Configuration Items. The CPCI is typically a computer program which is one of many elements in military, space, and/or civil systems. However, some of the required computer programs do not fall into this category and function off-line or completely separate from the operational system. There is then some leeway in the selection of which programs are to be identified as CPCI's. The designation of which computer programs are to be subjected to the full requirements of software configuration management is left to the discretion of the project office and the procuring agency. The designation is made on the basis of criticality, cost, and need for formal control, subject to the following guidelines:

1. At minimum, operational programs used in direct real-time support of a mission (or systems objectives) should be classified as a CPCI and/or further apportioned into subtier CPC's commensurate with the software work breakdown structure, program complexity, or program cost.
2. Certain nonreal-time or off-line programs, such as support and utility programs associated with a CPCI, or data processing programs used for telemetry reduction, may be designated as CPCI's if one of the following conditions applies:
   a. The status of the program directly affects project schedules.
   b. Changes to the configuration of the program directly affect the configuration of designated CPCI's.
   c. The programming effort requires a large expenditure of resources. In this case it may be deemed desirable to achieve a high level of management control through the use of configuration management procedures.

Although it may be decided through the previous process that all computer program development efforts are not designated as CPCI's, the elements of configuration management (e.g., technical documentation or change control) should be made applicable, although in moderation, whether the programs are identified as CPCI's or not.

Major Computer Program CPCI Restructuring. Consideration must also be given in the identification process to the degree that a program, performing the same general function, changes in systems use. In some cases, only data may change; in other cases slight program revisions may be required; in still other cases extensive modification may result. Hence, it is difficult to establish rigid rules to determine if a program change should be classified as a new CPCI or should be considered as a change to an existing CPCI. The following guidelines, however, should be followed in determining whether or not a program is to be reidentified as a new end item (that is, given a new identification number and defined by a new specification):

1. A program originally used may be modified and used without re-identification if:

a. Changes to its design are minor enough so that new design reviews are not required.

b. The existing software design specifications and related documentation can be modified by the use of specification change notices or minor revision.

c. The original program will not be used again.

2. A program should be reidentified if any of the following apply:

a. New design reviews are required.

b. New subfunctions requiring additional program modules to the CPCI are required.

c. A change of computing equipment causes a significant reprogramming effort.

d. A new specification is required to incorporate extensive modifications made over a period of time.

e. Slight changes are required, but the original version of the program will be used again. (Generic and concurrent versions of a basic program are an exception to this condition, and therefore the basic program does not require reidentification.)

21.2   SOFTWARE DEVELOPMENT DOCUMENTATION

It is highly desirable that the documentation structure for computer software merge hardware design methods of formal control (performance

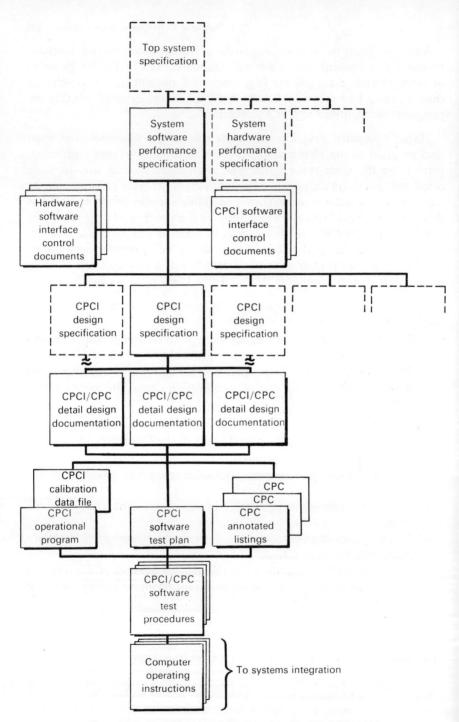

Figure 21.1 Typical computer software documentation structure.

specifications, interface specifications, test procedures, and so forth) with documents peculiar to software design (listings, detailed computer program flow diagrams, operating instructions, and so forth). This integrated approach allows existing hardware development management methods to be utilized for software development.

Objectives. It is intended that the computer software development documentation described here will permit achievement of the basic objectives listed below:

1. Provide project management visibility (that is, review, comment, coordinate, modify, and approve) of software conception, design implementation, validation, and configuration control activities.

2. Provide documentation of programming requirements, contents, and performance criteria at appropriate times during the development cycle for the effective coordination of program generators contributing to the computer program system.

3. Establish an orderly collection of design data that will ensure complete design documentation as the software system evolves rather than as a separate activity at the end of the development cycle.

4. Provide systems coverage for interfacing other programs and machines; that is, programs, computer, and/or functional devices.

Specific Documentation. The following specific document types will be prepared during the software development cycle (see Figure 21.1):

1. Systems software performance specification.
2. Systems software interface control documents.
3. CPCI design specifications.
4. CPCI/CPC (modules) detail design documentation.
5. Annotated listings.
6. Computer operating instructions.
7. Software test plans.
8. Software test procedures.

Systems Software Performance Specification. The systems software performance specification is considered to be the requirements baseline document for the software. It shall utilize standard specification format and define the requirements sufficiently for normal change control. The level of detail provided in this document should not unnecessarily constrain implementation but should thoroughly define each technical capability required. As a minimum, it should contain or reference the following data:

1. Specific allocated performance requirements, intra CPCI.

2. Input and output assumptions, constraints, and basic format requirements. (Top level system block diagrams showing all CPCI's and referencing applicable interface control documents are included.)

3. Required accuracy and/or timing limitations, tolerances, and capability alerts concerning critical areas of the program or interfaces with hardware or human factors.

4. Test and quality assurance requirements, responsibilities, and methods.

Systems Software Interface Control Documents. The systems software interface control documentation should be an integral part of the project interface control system. The documentation should include at least the following:

1. System input/output descriptions and formats.

2. Usage by function of all common data sources are described. Usage includes items either set or read from common data areas.

3. Inputs and outputs of the function will be defined and described; the equipment used for I/O^2 handling; and the frequency of data handling.

4. Interrupt priority system interfaces.

CPCI Design Specifications. These documents define the organization of the software elements (CPC's) and provide general implementation requirements for these elements. The specific data to be provided include:

1. Functional top-level flow charts of the overall CPCI, specifically showing the relationship of each CPC.

2. A concise explanation of the implementation shown in the flow charts.

3. Conventions, constraints, and general implementation ground rules.

4. Estimates of (and/or requirements placed upon) computer timing and core allocations including overlay budgets.

CPCI/CPC (Module) Detail Design Documentation. These documents contain the complete programming plan for each software CPC (module). They ensure that the following minimum requirements are met prior to committing excessive amounts of machine time for checkout:

1. The programmer/analyst has sufficient functional information to complete the detailed program design and coding.

2. The technically cognizant subproject(s) officers, whose software requirements are being implemented, can determine that their needs are met by the proposed design.

These documents are a direct outgrowth and extension of the CPCI design specification. They need not repeat data that are contained in the CPCI design specification. It is intended that these documents be the product of a team (engineer/programmer) effort and functionally represent the actual software to be ultimately delivered (after software system test). It is not

---

^{2} I/O—Input/Output.

intended that the documents contain programming detail that is best provided in the annotated listing.

The specific contents of these documents include the following:

1. *Functional flow chart*, which describes, in a functional manner and in engineering terms, what is being accomplished at each step in the program. The connection to the top-level flow chart(s) is identified as well as the connections to other flow charts. Where the task being performed is largely mathematical, the flow chart may consist of a simple box with reference to another written document. It is intended that the flow chart represent the primary design communication method with written specifications, documents, and so forth, used only when appropriate and then indexed by the flow chart. The level of detail required to show every functional branch is provided. Utility programs (multiple-use routines) should be consistently shown by conventions and agreements established in the top CPCI design specification or by reference to other program design documents.

2. *Functional descriptions*, which provide a concise explanation of the program's function and are used to support the interpretation and use of the related flow chart.

3. *Detailed reference material*, data which are considered to be part of the program design document by being specifically referenced in the flow chart. Where the material referenced in the flow is a controlled source document, it need not be enclosed but should be thoroughly identified. Specific types of data should include:

    a. Accuracy and limitations of logic processes.

    b. Hardware requirements and/or limitations.

    c. Limits to volume rate of input and/or output data.

    d. Initial condition requirements both for hardware and software.

    e. Execution time estimates where not specified as requirements on the flow chart and where they affect program performance.

    f. Detailed procedural methods and formats of manual data entry.

    g. Input and output definitions and format.

    h. Size estimates (words, bytes, and so forth).

Annotated Listings.   Annotated listings serve as the primary detailed documentation to control and define the actual software product. The listings include or have appended at least the following:

1. *Identification*, the name mnemonic plus working deck information (ID) and modification (MOD) program author(s) and date, as well as the calling sequence and elements.

2. *Storage requirements*, the number of memory cells required, with a distinction made between program and storage area. A storage scheme is appropriate here.

3. *Timing*, a brief description of the time required for execution and identification of any critical time dependencies, such as input/output equipment time.

21.3 COMPUTER SOFTWARE DEVELOPMENT PHASES

The major software development phases and activities are diagrammed by Figure 21.2 and are discussed in the following material.

Conceptual Phase

Requirements Formulation (Preliminary and Engineering Design). The requirements formulation activities develop the top level and detailed software requirements through active coordination, analysis, review, and implementation planning. Because essentially all major project activities normally have requirements that may influence software, this task requires thorough interface coordination. This phase is considered to be critical in that errors at this level can have much larger impact on cost and schedule than those that may occur later in the development cycle. Completion of

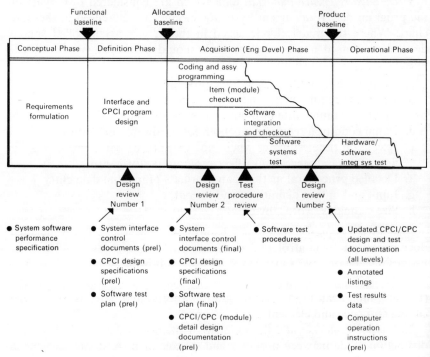

Figure 21.2 Computer software development phases.

this phase, which results in systems software performance specifications and preliminary interface control specifications, heavily influences all subsequent development tasks.

Definition Phase

Interface and CPCI Program Design. The CPCI program design defines the functions to be performed in each CPCI which will meet the requirements of the systems software performance specification defined in the above. The general functional methods to be used, input, output, interfaces, and data characteristics are determined. In addition, guidelines for generality and expansion required in the program are provided and designation of previously prepared subroutines or programs which are used intact or modified for use in the system are identified. The preliminary interface control documents prepared during requirements formulation are updated and formalized.

CPCI/CPC (Module) Detail Design. The detail design phase determines the detailed functional methods of implementation to be used. These include constants, limits, input and output formats, calling sequences, logic flow, interrupts, addresses, and general core and timing constraints.

The functional logic flow for each module is prepared by the programmer analyst with all functional branch points, processing (subroutine call) blocks, data tables, core allocations and overlays, entry points, and error recovery routines. Specific information is provided concerning external storage to be used in the system, with formats and I/O methods to be used including libraries and cataloging conventions.

Acquisition Phase

Coding and Assembly (Programming). This activity develops the individual elements of software. It includes acquiring and modifying previously written (library) routines, writing new routines, loading, assembling, and debugging the software module. All programs should be liberally sprinkled with comments which define branch points, meaning or definition of names used, and action or function of processing (called) blocks. Each routine is then tested to ensure proper logic manipulation. The required sublevel activities are the following:

1. *Library assembly*, which involves acquiring previously prepared programs or subprograms to be used in the system. This includes identity of "built-in" or hardware vendor supplied routines. Each routine is analyzed for suitability, size, and so forth; changes required are identified and defining documents modified.

2. *Programming and debugging,* which deal with making all changes to the library routines and writing all new software modules, along with any unit control software required to link the modules used. Programs are placed on punched cards, and/or magnetic tape, loaded, compiled (assembled), and all errors corrected.

CPC Checkout (*Module Checkout*).   The smallest runable modules are tested to exercise all branches and processes in the item. Program listing and test results are available at the termination of this phase. Both the system analyst and programmer analyst are active in this phase, and operation is checked against predicted output. Error recovery routines are exercised and checked.

Software Integration and Checkout.   All CPC programs at the CPCI level are brought together, linked, and exercised as a unit. Each step of the integration requires participation by the programmer analysts and system analysts that have been involved in the development effort. Checkout methods previously used for item checkout are exercised again to provide a first check in the integration. New test cases are provided by the analysts and engineers to provide integrated data to test the subsystem. Timing and core budget information is recorded and all problems are identified and corrected. New modules, corrected modules, and so forth, as required, are assembled, checked, and integrated. Activities under this phase are the following:

1. *Software CPCI integration* is the phase which assembles groups of CPC's (modules) together that are used together in a particular phase of systems/mission accomplishment. These are checked out with test data and/or hardware as required.

2. *Software CPCI checkout* occurs when the full software CPCI is loaded and actual service is attempted for each external hardware interface based on interrupts, input, processing, and output. Results at all levels are recorded.

3. *Software System integration* brings all CPCI's together that will operate under the control of one computer executive. These are checked out with test data and unit hardware including other computer systems as necessary. The completion of this phase results in a running software system which is ready to undergo formal software system testing. As a consequence, formal software change control must be exercised from this point onward.

Software System Test.   Software system tests are performed to released test procedures and operating instructions with the full software system loaded, hardware connected (where necessary), and simulated data (where necessary) available. The tests operate as closely as possible to the real-time system, with sufficiently realistic and emergency work loads imposed as the engineers, analysts, and programmers consider necessary for software system checkout. All inputs, processes, and outputs are recorded and checked in detail for any logic, mathematical, control, format, timing, limit, or other errors in the system. Extreme care is taken to ensure that change control documentation is maintained at all times. (See Section 21.5.) The completion of this phase leads directly into the initiation of the final real-time hardware/software integration test.

Hardware/Software System Integration.   The formal software system test is repeated to include *all* hardware and software performing in the actual environment in which the system will operate with actual hardware linkages intact. The first tests may simulate hardware components not yet available, but all hardware interfaces must be tested before completion. The intent of this activity is to exercise the software, computers, and all related external hardware with as close to an operational environment as possible. It provides a last check on all formats, processes, control, and integration provided in the computer hardware and software before the operational phase. It also provides timing, computer workload, and final core budget information to ensure the operational readiness of the system. Completion of this phase of testing constitutes final acceptance by the customer of the CPCI(s).

Operational Phase

Special demonstrative testing of the system or its major modifications may be required during the operational phase.

Software Design Reviews

The design review definitions outlined here provide the major schedule events for the software development effort and focus technical and management attention on that effort at critical design phases. Standard design review procedures shall be followed wherever possible.

Design Review No. 1   Preliminary Design Review.   The purpose of this review is to examine the results of the software requirements definition, CPCI program design implementation, and the development planning effort. The contents of the data package are defined to include at least the following:

1. Systems software performance specification.
2. CPCI program design specifications (preliminary).
3. Systems interface control documents (preliminary).
4. Test plan (preliminary).

Design Rewiew No. 2 Detailed Design Review(s). The purpose of these reviews is to verify the adequacy of the detailed program design effort. The review package should contain at least the following:

1. Updated CPCI program design specifications.
2. CPCI/CPC detail design documentation for the software element(s) being reviewed.
3. Updated systems interface control documents.
4. Final test plan.

Test Procedure Review. The purpose of this review is to verify the adequacy of the CPCI/CPC test procedures against test plans and design requirements.

Design Review No. 3 Software Product Design Review. The purpose of this review is to examine the results of the program integration, checkout, formal software test, and initial documentation effort. The review package should contain or reference at least the following:

1. Updated CPCI/CPC design documentation (all levels).
2. Annotated lists and test result data.
3. Computer operating instructions (preliminary).
4. Requirements versus capabilities check list.

21.4 COMPUTER SOFTWARE CHANGE CONTROL

As a configuration management function, change control is defined as a system by which proposed changes may be analyzed and coordinated with all impacted activities prior to authorizing or rejecting the change. If authorized, the change implementation is verified by systematically comparing by visual (for documentation) and by test (for function) the changed product.

Change Control Organization. The execution of configuration management will be the responsibility of the manager of the project configuration management office (CMO).

The CMO Manager or, depending on project scope, his special designated representative (software program configuration manager—SPCM) for software controls, is not totally responsible for detailed technical decisions but nevertheless must have a firm grasp of the software's technical factors, organizational responsibilities, customer requirements, current baseline configurations, test requirements, and accepted configuration management procedures. The SPCM supports the project configuration control board (CCB).

Software Development Change Control.   During the development phase before the start of the formal software system, the major formal configuration management control is provided by systems performance specifications, design specifications, and interface control documentation. The documents, when released, require CCB approval of the software change request (SCR) (see Figure 21.3).

Software Product Change Control.   The two major problems in the long term reliability and operability of software are obsolescence of documentation and errors introduced by modifications of the *software product* that were not properly evaluated, authorized, and/or verified. If these two problems are allowed to reinforce each other, the end result can be chaotic. Therefore, all changes subsequent to software integration and checkout must be carefully controlled by SCR with respect to the *software products* as well as their authorizing specifications. Accordingly, the following requirements for the control of software products are emphasized (see Figure 21.4):

1. The software product is defined as the operational program that is released for formal software test. It will have completed checkout and been reviewed for overall adequacy.

2. The software product is released for software test in the form of a symbolic deck and, where required, is on magnetic tape, disk, and so forth. An acceptable deck is defined as one properly sequenced and identified and suitable for listing and assembly: hence, one able to furnish proper results from tests defined in the formal test procedures.

3. Formal change control of the annotated listings and test procedures is established before Formal Software Tests.

4. The software program master and backup editions (cards, tapes, disks, annotated lists, and so forth) are impounded by quality assurance and stored within the software data center. The program master is released for use (i.e., retest) only with quality assurance surveillance.

5. Because of the software product's value and importance, it shall be stored by quality assurance in a secure area. This version of the product is known as the software product "master file" to differentiate it from working decks, tapes, disk-files, and so forth, that may contain identical information. Extreme care will be used to ensure the master file is not changed without authorization and is protected against accidental damage.

6. Approved test procedures will be formally released for use by quality assurance during test surveillance. Certification of test results (test record sheet) by quality assurance is required.

7. Proposed changes to the operational program resulting from identification of new requirements or logic errors will be programmed and checked

Software Change Request	Date:	SCR ___

Project:	Project Task/Subtask ID:	Originator's Name:

Priority: ☐ Critical Request ☐ Routine Request Change Category: ☐ New Requirement Error ☐ Corrective Action ☐ Improvement Suggestion

Originator Section—Note: only one item per report

Title:

Item(s) Affected
☐ Total System ☐ Software Only
☐ CPCI ☐ Hardware
☐ CPC ☐ Other: ___

ID Numbers of Items Affected:

Purpose and Salient Features of Suggestion or Problem Highlights:

Description of Modification:

Compatibility with Existing Software and Hardware Interfaces:

Work Package or Task Manager Evaluation Section

Analysis Remarks

Estimated Workload Data

System Analyst ___ MHS   Machine Time ___ Hrs   ☐ Approved
                                                               ☐ Rejected

Program Analyst ___ MHS   Time in Weeks

Documentation ___ MHS   to Complete ___   CCB Chairman | Date:

Other (specify) ___ MHS   Estimated Start Date: ___

Analyzed by:	Date:	Authorized by: (W/P or task Manager)	Date:

Figure 21.3 Software change request (SCR).

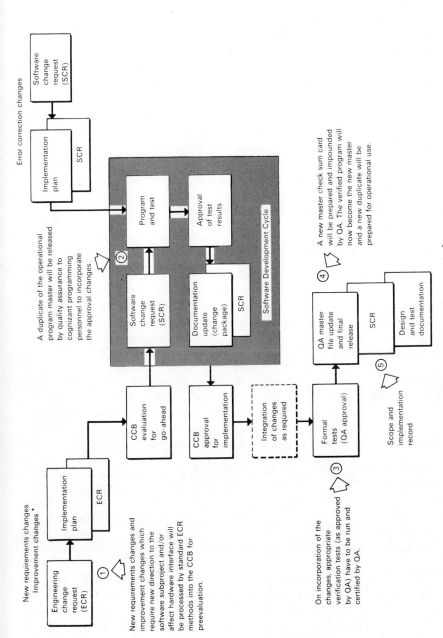

Figure 21.4 Software change control.

out using a scratch program provided for that purpose. Upon receipt of authorization from the CCB to modify the computer program, the master and backup editions will be released by quality assurance to cognizant software programmer personnel to incorporate the approved changes.

8. The retest program must be reviewed by quality assurance to verify that the retest requirements are just as exhaustive as the testing of the original software.

9. Except for minor documentation corrections or clarifications, change requests to the software program shall be entered on a software change request (SCR) form and processed to the CMO.

10. The expediting of changes arising from formal software tests and requiring immediate action may be enacted without SCR documentation, providing the aforegoing quality assurance provisions are maintained. However, subsequent back-up SCR documentation is required.

21.5 SOFTWARE PRODUCT IDENTIFICATION

As a configuration management function, product identification has been defined as the means by which end products are to be identified and associated with their respective software configuration documentation. The following requirements are presented for the purpose of establishing a consistent and unambiguous method of identifying the software product and all changes to it.

Computer Program Configuration Item Identifiers. Each individual CPCI requires a unique reference number which, once assigned, is not altered except under the conditions set down under section 21.1, paragraph entitled "Major Computer Program CPCI Restructuring." The CPCI number consists of an alphanumeric six digit identifier.

Part Numbers (Computer Materials). A part number is assigned for each identified "part" within the CPCI. A part is defined as the card decks, magnetic tapes, or the like, in which the program is contained. A physical part as defined above may in many cases correspond to a logical part, a computer program component (CPC). Identification requirements for computer program parts are described in the following paragraphs.

Card Deck Identification. Each card deck containing a CPC is given a header card and a trailer card. Each header and trailer card contains the name and number of the CPCI and the name and part number of the CPC module. These cards are treated as comment cards by the assembler or compiler, that is, they have no program functions other than identifying a particular card deck.

Each card within a card deck is marked by a *printed* CPC part number and a punched card identification/sequence number. Each CPC card deck is also color coded for easy identification. The band or case of each subprogram deck is marked by both the CPCI number and CPC or logical module part numbers.

The construction of the card identification/sequence number is illustrated in the following:

1. *Fortran Cards*

Each CPC or logical module of coding is identified in the following manner:

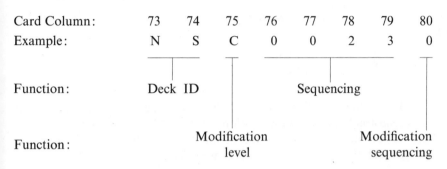

Card Column:	73	74	75	76	77	78	79	80
Example:	N	S	C	0	0	2	3	0

Function: Deck ID   Sequencing

Function: Modification level   Modification sequencing

a. *Field 1 . . . card columns 73, 74*

These two columns identify the subroutine and are punched the same on every card of the deck. Characters 0–9, A–Z are allowed, yielding 36^2 combinations. Although these two characters do not necessarily identify the function or relate to the subroutine name, it is not necessary that the sequencing satisfy these conditions.

b. *Field 2 . . . card column 75*

This column will identify the modification level of a deck. Initially each card of the deck will be punched with an "A" in column 75 signifying version "A" of the routine. When the first modification is made, each card in the deck will be punched with a "B" in column 75; on the second modification a "C" will be punched, and so on. The second card of each deck will be a comment card containing the document number associated with the listing for this deck and the current version number for easy identification. As many as 26 sets of modifications may be identified in this manner.

c. *Field 3 . . . card columns 76–79*

These columns are used for numerical sequencing. Decks of up to 9999 cards can be sequenced uniquely. The first card of the deck (the subroutine name card) should begin with 0001.

d. *Field 4. . . card column* 80

This column will be 0 on all quality assurance controlled decks. It will only be used by the programmer as an aid in sequencing modifications during checkout of a new version of a subroutine. Column 80 will allow up to 9 inserts between existing cards in the controlled deck.

Example: If two cards are to be inserted between cards

NSC 00230
NSC 00240

the inserts would appear as

NSC 00230
NSD 00231
NSD 00232
NSC 00240

When the modifications are validated, a new deck will be issued reflecting the new modification level ("D" in the above example) in column 75 and new sequence numbers in columns 76–79. Column 80 will always contain 0 in the master decks.

2. *Assembly Language Cards*

The method of identifying the cards which represent subroutines of logical modules of coding for programs coded in assembly language is as shown below. This method is the same as described in (a) above.

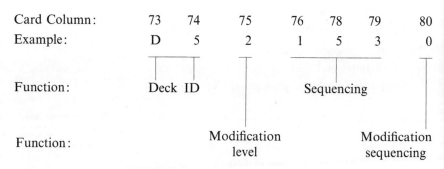

Card Column:	73	74	75	76	78	79	80
Example:	D	5	2	1	5	3	0

Function: Deck ID   Sequencing

Function: Modification   Modification
           level   sequencing

Tapes and Disk Identification. Magnetic tapes and/or disks used as mediums for storing operational software and which are to be controlled by the project quality assurance are identified as follows:

1. The configuration management office issues a tape or disk identification number and records, in a file created for such purpose, the identification and modification level of all items included within the storage medium.

2. The storage medium is labeled with the issued identification number.

Program Listings. Annotated listings of the operational programs which are to be controlled by the project quality assurance are identified as follows:

1. The configuration management office issues a listing identification number and records, in a file created for such purpose, the identification and modification level of all CPC's or logical modules of coding which are included within the listing.
2. The listing is labeled with the issued identification number.
3. In addition, the listing contents automatically include the identification of each card which was used in the compilation.

21.6   QUALITY ASSURANCE VALIDATION OF SOFTWARE PROGRAMS AND PROGRAM CHANGES

As specified earlier, major program changes are presented to the CCB for approval prior to formal implementation. In addition, the master computer program (software products) is stored in the data center and controlled by quality assurance. Accordingly, all software products that are under quality assurance custody are restricted to the ground rules set forth in the following. The basic concept is illustrated in Figure 21.4.

Check Sum Requirements. The application of check sum routines is to perform an arithmetic sum on computer software to indicate a go/no-go condition to verify the configuration of software programs. This routine will allow quality assurance to be kept current on all software program changes because program changes affect the sum directly. The provisions of the check sum technique are as follows:

1. A check sum routine is provided by software personnel for each module or logical segment of the computer program. The check sum routine provides a method in which programs are called into core from an absolute library and perform an arithmetic summing process on the memory cells resulting in a full bit representation of a check sum number as appropriate to the computer configuration so programmed. This in effect results in comparing the master file program against any current programs that are stored in the computer. (Caution should be taken that compiler or editor variations do not compromise the check sum process.) The computer stops and indicates by print-out if a no-go condition occurs. A separate debug routine is utilized to indicate by print-out the differences occurring between programs.
2. Quality assurance provides surveillance during a debug routine to establish the cause and area of a check sum failure for the purpose of documenting failures. Quality assurance writes a failure report for all such failures.

3. Key cards, which are developed as a result of the check sum routine, are issued to and controlled by quality assurance. Revisions to programs require new key cards for these revisions. Quality assurance requires verification that the revised key card confirms the revision.

4. A check sum is required by quality assurance prior to running all formal software programs.

5. Software products; that is, cards, disks, tapes, and listings in the data center, will be removed from the data center for performance of tests by quality assurance personnel only.

Quality Assurance Test Monitoring and Acceptance. Formal configuration control of the annotated listings, computer program decks, tapes, and so on, and test plans/procedures is initiated prior to formal software systems tests.

Formal quality assurance procedures are also initiated at this time. In addition to the check sum objectives specified, the following detailed requirements are necessary:

1. Prior to approval of all software test procedures, quality assurance verifies that all requirements dictated in the controlling test plan are adhered to in the test procedure.

2. Quality assurance must have possession of a copy of the software test procedure, stamped "Authorized Test Copy," along with all necessary approval signatures prior to test performance.

3. Authorized test change records (TCR's), test record sheets (TRS's), and procedure change orders (PCO's) may be used to modify criteria of an approved test procedure. All changes to formal software test procedures are processed through the CCB.

4. Quality assurance surveillance of all formal software tests and certification of all test results are requirements.

5. Quality assurance maintains a log on software for recording all software failures and problem areas of concern occurring during formal software tests.

6. A back-up tape on all software programs is retained in the event that master files are destroyed. If the system configuration utilizes magnetic tapes and not disks, controls can be accomplished using a three tape rotation system controlling delta changes and back-up tapes.

7. Quality assurance verifies the test setup before the beginning of testing and documents deviations to procedural requirements. A configuration listing of all hardware involved in the test is provided for each test data package by quality assurance, supported by the CMO. This list includes part number, serial number, revision letter, and rack location.

8. The test data package is comprised of data sheets, computer print-outs, and a test configuration list. Software test data packages are subject to test review board approval.

9. Quality assurance retains, one change back, all updated master card decks and creates a "dead file" three changes back for annotated listings.

Preshipment Software Inspection. At the conclusion of factory integration testing, a preshipment inspection of the software component is performed. The following documentation and software products are carefully identified and itemized on the shipping record.

1. Copy of computer log.

2. Copy of output print-out, selected in accordance with test procedure requirements.

3. Input and control data, on cards and tapes.

4. Computer program source deck listings end item data.

5. Computer program source decks and end item object code decks on cards.

6. List of shortages.

Chapter 22

THE CONTEMPORARY
AEROSPACE SCENE

22.1 DOD AND INDUSTRY MANAGEMENT DEVELOPMENTS

Chapter 1 discussed the difficult task undertaken in 1966 by DOD in concert with industry to alleviate the problem of the increasing proliferation of divergent and incompatible management systems imposed upon industry by the various functional areas within OSD (Office of the Secretary of Defense), the Army, Navy, and Air Force, and other government agencies, such as NASA. Although industry was instrumental in initiating the "generic" management concepts such as AFSCM 375-1, PERT/Cost, and so forth, management off-shoots of the basic system were becoming too much of a good thing. By 1965, the government was "holding hands" with industry every step of the way, tending to curtail its initiative and flexibility. Contractors were pleading for disengagement, asking the government to stand back and let them do the job they were being paid to do.

The very organizational structure and the attendant echelons of authority within the DOD seemed inherently to force proliferation of data and embroidered implementation directions. Each succeeding level of management tended to interpret and detail the "bosses'" desires. If the initial pronouncement was detailed as to what, when, why, where, and how, it was axiomatic that there would be an elaboration in some respect at each level of implementation. Therefore the stage setting in 1965 was one of a complex interrelationship of a greatly expanding number of management systems of all kinds, emanating from different functional arms of the DOD and the services, in a variety of time phasings, and all impacting on the typically multicustomer aerospace company. Consequently industry pressed for a program favoring management system disengagement between DOD and industry and for a single point within DOD to control and evaluate new management systems proposed by the services.

22.2   DISENGAGEMENT

"Tell me what you want but not how to do it" was industry's repeated refrain to DOD in early 1965, a plea for disengagement. The objectives of disengagement pressed by Industry during 1965 dealt with reduction of the number of control actions, amount of *detailed* government management, detail of management visibility (reporting) media, and elimination of control actions inconsistent with any form of fixed-price or incentive contracting. With respect to this latter point, it appeared a contradiction to have the government heavily review *and approve* contractor actions without seriously weakening contract incentives and warranties. It was the opinion at that time that if the government chose to exercise the right to take *detailed* control actions, it must appropriately share in the success and failure of contractor performance with respect to incentives and fixed-price limits. On the other side of the scale, the government obviously could not relinquish any portion of its responsibility for the stewardship of appropriated funds or technical and economic commitments. Therefore industry and the government found it necessary to seek new tools to allow the government to manage-by-exception yet still react to immediate decision conditions based on management information provided by the then existing, more scheduled, review and approval process. This may best be termed "responsive visibility."

22.3   MANAGEMENT SYSTEMS CONTROL

During September 1965 an ad hoc group, Systems Management Analysis Group (SMAG), was formed within the Aerospace Industries Association of America (AIAA) to investigate the increasing proliferation of government management systems. In the initial SMAG report in November 1965 the problem was identified as the most serious operational problem facing industry. Because of its serious nature, it was decided by the Executive AIAA Council that the scope and findings of the SMAG project would be discussed initially at only the highest levels of the OSD and industry. A final SMAG report which outlined industry's major areas of concern was presented to DOD on May 12, 1966.[1]

The problems highlighted were directed at the conflicts between management systems, the mating of appropriate systems to be used in a given case with the type of contract selected, and the need to tailor the degree of management to the complexity of the program involved. It was also urged that each new management system be carefully examined before its adoption to assure its consistency with other systems, to assure its consistency with the

[1] Karl G. Harr, Jr., AIAA President, "Management Systems," *AIAA Letter*, December 7, 1966.

overall body of DOD policy, and to assure that the new system is, in fact, worthwhile when considered in light of the expense involved in its application.

To set in motion resolution of the overall problem, the AIAA recommended that OSD do the following:

1. Establish a single office in OSD with direct operational responsibility for all management systems and counterpart offices in each service.

2. Establish and empower a small management team comprised of OSD and industry to develop a master plan and schedule for top management approval.

3. Through use of CODSIA[2] initiate joint DOD-industry effort.

DOD Resource Management, DODD 7000.1

The force of these arguments directly contributed to the release of DOD Directive 7000.1, August 27, 1966, concerning management systems; this document established the Assistant Secretary of Defense, Comptroller, as custodian of all program management systems. Any system related directly to financial control or reporting was to be established by the Comptroller, while others were subject to his approval. Thus a central source for management systems control was established, as industry had requested. Uniformity to the extent appropriate is the goal in the application of *existing* systems, while new concepts must, first, be necessary, and second, be *proven* on a small scale before implementation.

System/Project Management, DODD 5010.14

During this same period, another major influence on DOD centralized management systems control policy was the release of DOD Directive 5010.14 on system/project management. This directive established policy for assignment of system/project management responsibility to a designated individual. This individual has centralized management authority which enables him to cut across all the interservices or DOD component authorities when necessary. Projects, or tasks within a project, are not divided between the services in a manner which would give individual services a license to manage on their own; rather, authority is vested in one individual within the services assigned. Thus, when more than one service is participating, each reports centrally to the system/project manager.

DOD Management Control Systems Development and Selection

With the release of DODD 7000.1 and DODD 5010.14, the OSD-industry objective for centralized (organizational) management systems control

---

[2] CODSIA, Council of Defense and Space Industries Associations.

within DOD was formalized. Two additional objectives of the DOD Management Systems Control Plan of 1966, which specifically dealt with the controls for the actual development and selection of new management systems, were acted upon by the June 1968 release of two DOD instructions: DOD Instruction 7000.6, "Development of Management Control Systems for Use in the Acquisition Process," and DOD Instruction 7000.7, "Selection and Application of Management Control Systems in the Acquisition Process."

DOD Instruction 7000.6, Management Control Systems Development. The purpose of DOD Instruction 7000.6 is to establish a procedure to be followed by all DOD components when there is thought to be a need for a new management control system. The instruction outlines the review and approval process that will be required and the justification that must be submitted with the request for approval. It sets forth the criteria for the kinds of management information to be brought under this control and specifically excludes technical specifications and standards and procurement policies.

DOD Instruction 7000.7, Selection of Management Control Systems. DOD Instruction 7000.7 provides a formal procedure for the selection of management systems to be applied to contracts, based on the nature of the acquisition including a review and approval of the systems selected for application.

As part of the implementation of this instruction, the DOD Comptroller's Office has published a management systems control list (MSCL) containing some 1500 documents identified as "management systems" by each of the military departments and the Defense Supply Agency. The MSCL list contains MIL specifications, MIL standards, ASPR regulations, service regulations and data items as well as management systems. It also contains some systems that are only used internally and impose no requirements on contractors.

DOD with industry participation, will screen this list with the objective of eliminating duplications and consolidating overlapping documents. The approved documents in each functional classification (technical data, logistics, quality assurance, and so on) will be transferred to the authorized management systems control list (AMSCL). When the screening process has been completed, the AMSCL will contain only those items that are classified as management systems and will be the only management systems that can be put on contract.

A DOD project is also underway to establish similar authorized lists for specifications and standards. Because of the manner in which these documents have been written over the years, many of them will have to be reworked before being classified into one of the three authorized control lists: management systems, specifications, or standards.

22.4   TREND TOWARD "RESPONSIVE VISIBILITY" AND DISENGAGEMENT[3]

As we have shown, both industry and government have been making considerable progress in their efforts to control the undergrowth of management systems by some judicious thinning.

With respect to disengagement, on the other hand, the ideal degree of active government participation in a contractor's systems management is a delicate balance, not as easily achieved; though there are indications that some noteworthy progress is being made in addition to DOD's management systems control program.

As suggested earlier, the ideal degree of government engagement can be achieved by what might be called "responsive visibility," or assuring that the government gets sufficient information and control to see program progress and problems clearly at any given moment, thereby assuring industry that the government will step in effectively with its own management resources only where and when it becomes apparent that the contractor's effort is inadequate and headed for trouble.

Historically, the degree of government active engagement in the internal management of defense industry has fluctuated with changes in the nature of defense material, in the urgency of hardware needs, and in government philosophies and techniques of procurement.

The trend during the 1950s and very early 1960s, for a number of reasons, was toward increasingly deeper government engagement. At that time revolutionary and extremely complex ballistic weapon and space systems were being developed under the pressures of urgent priorities. Both government and industry were pioneering wholly unfamiliar territory in high-risk systems of such complexity that development costs were too great to be borne by private industry. There was, of necessity, a heavy reliance on sole source procurement, since no truly competitive capabilities had yet been developed within industry for these types of systems. In 1961, for instance, 85 percent of the Air Force's awards were noncompetitive; 46 percent were cost-plus-fixed-fee.

The amount of autonomy given to industry, as well as industry's profit, is related directly to the degree of risk which industry assumes and to the

[3] Summarized, in part, from Brigadier General Daniel E. Riley, USAF, Commander, Air Force Contract Management Division, Air Force Systems Command, "The Government's Role in Minding It's Contractor's Business," *Defense Industry Bulletin*, Vol. 4, No. 4, April 1968.

element of competition in the procurement atmosphere. With both at a low ebb, government intervention with contractor management tended to increase.

Another aspect of early space-age procurement tending to stunt development and improvement of industry's internal systems management capabilities was the government practice of "piecemeal" procurement. Because it was difficult, if not impossible, to estimate in advance the exact total performance and cost of the systems, it became the normal practice to award only the development work at the outset of a program. Unless the contractor was a spectacular failure in this phase of the program, he was practically assured of the follow-on procurement, with no commitment concerning ability to control costs, assure performance, or meet schedules. The alternative, selection of a new contractor, meant duplication of the greatest part of the original development costs. The government tended to overcompensate for the known but unavoidable shortcoming of this procurement method by assuming an unusually active role in monitoring industry's internal management.

Contributing still further to the tendency toward overcontrol by the government was the mid-century revolution in the tools and techniques of data processing and storage. Helpful as the new computer capabilities were, they tended to encourage the proliferation of new reporting requirements. The government became involved in ever more detailed surveillance of industry management.

By the early 1960s, however, the procurement atmosphere, particularly in the Air Force, was beginning to undergo marked changes. After almost a decade of highly concentrated space-age experience, an invaluable working knowledge of the realities of space-age technology had been acquired. Greater attention was gradually concentrated on integrated defense planning and management, more effective use of resources, and improvement of the acquisition process and the general climate of the government-industry working partnership.

One angle of approach was the drastic reduction of cost-plus-fixed-fee contracts. A substitute was the negotiated fixed-fee contract, based on weighted guidelines which took into consideration the element of risk for the contractor and the contractor resources, capital, and skills required.

The usefulness of the fixed-fee contract has been further enhanced by the addition of incentive arrangements which reward the contractor for improving upon specified hardware performance, cost, or delivery schedule, and penalize him for failure to meet performance, cost, or schedule objectives originally established. This prospect of higher profits and threat of loss constitutes a most effective incentive to industry to put forth a maximum effort in good management.

The successful shift from cost-plus to fixed-fee with incentive contracting is amply attested by the record. In FY 1962 46.9 percent of all Air Force contracts were cost-plus-fixed-fee. By FY 1967 the percentage had dropped to 5.1.

Improved source selection in awarding contracts to industry is also proving to be a promising approach to a healthier balance in the government-industry management relationship. An armed services procurement regulation revision of June 1, 1965, sets forth requirements for an exhaustive precontract investigation of company capabilities, as a measure to reduce the necessity for remedial government intervention at a later date. In addition, the contractor performance evaluation report, inaugurated in 1963, provides a continuing semiannual evaluation of performance on certain contracts. This report provides a long-term incentive to contractors by creating, within the government, a "memory" of contractor performance and a means for considering this record in future source selections and negotiations.

An additional development during the late 1960's, which was directed toward an improved management balance, was the trend toward total package and life-cycle procurement. One of the basic principles of total package procurement is a high degree of disengagement of control over contractors after a long-term program has been established under competitive conditions. Total package procurement is the antithesis of the piecemeal approach. It is procurement through one-time open competition of a maximum number of elements of a system, throughout engineering, development, production, maintenance, and so forth.

Though it does have certain inherent limitations and disadvantages, total package procurement was intended to put the competitive muscle tone back into government procurement. It was to integrate and simplify the procurement process as a whole; to get the contractor back into business on his own, permitting maximum government disengagement from his internal management.

All of these trend makers of recent years—the strengthening of the competitive element in contracting, improved source selection, tightening control of systems management and reporting requirements and techniques, the introduction of new procurement methods such as the total package concept—gave promise of progressively improving government-industry working relationships.

Blue Ribbon Defense Panel

With all the potential and the promise offered by the management evolution described thus far, the present aerospace scene—troubled by the controversial C-5A super-cargo plane, the Cheyenne helicopter, and the

F-111 fighter plane—reaffirms that the ideal degree of active government participation in industry's management is a delicate balance which is not easily achieved and is subject to shifting philosophies and national emergencies.

With defense procurement of major programs burdened by cost overruns and technical problems, President Nixon chartered a Blue Ribbon Panel[4] in July 1969 to study and report on the organization and management of the defense department. The panel took particular aim at total package procurement: "There should be no attempt to package the development and production of a plane or a weapons system and to project all of the costs over a seven-to-ten year period. Development should go step-by-step, checking the milestones where we can increase or decrease the buy. The project should be checked each step of the way, and production not to begin until there is no question at the equipment's reliability." The panel recommended a return to the fly-before-you-buy concept of more prototypes, more test hardware and less reliance on paper studies.

Although the panel's year-long study covered many areas, the following is a summary of their findings on the acquisition process of aerospace or other systems:

"A new development policy for weapons systems and other hardware should be formulated and promulgated to cause a reduction of technical risks through demonstrated hardware before full-scale development, and to provide the needed flexibility in acquisition strategies. The new policy should provide for:

1. Exploratory and advanced development of selected subsystems and components independent of the development of weapon systems.

2. The use of government laboratories and contractors to develop selected subsystems and components on a long-term level of effort basis.

3. More use of competitive prototypes and less reliance on paper studies.

4. Selected lengthening of production schedules, keeping the system in production over a greater period of time.

5. A general rule against concurrent development and production efforts, with the production decision deferred until successful demonstration of developmental prototypes.

6. Continued trade-off between new weapon systems and modifications to existing weapon systems currently in production.

---

[4] The entire report to the President and the Secretary of Defense on the Department of Defense by the Blue Ribbon Defense Panel, July 1, 1970, is available by mail from the Superintendent of Documents, U.S. Government Printing Office, Washington, D.C. 20402; price $2.25; order number D1.2/B62/970.

7. Stricter limitations of elements of systems to essentials to eliminate "gold-plating."

8. Flexibility in selecting type of contract most appropriate for development and the assessment of the technical risks involved.

9. Flexibility in the application of a requirement for formal contract definition, in recognition of its inapplicability to many developments.

10. Assurance of such matters as maintainability, reliability, etc., by means other than detailed documentation by contractors as a part of design proposals.

11. Appropriate planning early in the development cycle for subsequent test and evaluation, and effective transition to the test and evaluation phase.

12. A prohibition of total package procurement."

Therefore, at this writing, the debate, in part, is between total package versus piecemeal procurement. With regard to the return to piecemeal procurement and in defense of total package procurement, perhaps we can now afford to be more painstaking in the development of new weapons as we enter 1970, especially since the pressure to match Russian military development is less intense than it was a decade ago. In any case, it is not the intention of this book to take positions. However, it should be pointed out that the decentralization versus retrenchment of management controls as a function of time is as commonplace to the internal policies of the smallest contractor as it is to the highest management philosophies of the Pentagon or NASA. It is as inevitable as the tide.

22.5   THE *NEW* CONFIGURATION MANAGEMENT

It remains for us at this point to set the new DOD configuration management policy and standards (Figure 22.1) against the background of contemporary changes in DOD management discussed thusfar—disengagement, uniform management systems controls, responsive visibility, and so on—and examine their influence on the new standards.

Before doing so we identify the scope of the new configuration management authority documents and their address to traditional areas:

1. Authoritative policy and implementation guidance. (DOD Directive 5010.19 and DOD Instruction 5010.21.)

2. Criteria for selecting specification types for describing item functional and physical characteristics, and guidance for in-house or contractual preparation of these specifications. (MIL-S-83490 and MIL-STD-490.)

3. Criteria and uniform practices for proposing, justifying and approving engineering changes, waivers and deviations, and methods for their implementation. (MIL-STD-480 and MIL-STD-481.)

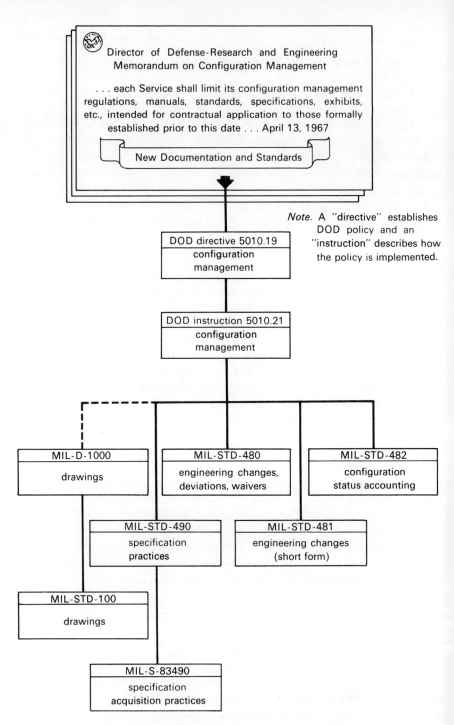

Figure 22.1 DOD configuration management standards.

4. A comprehensive listing of standard data elements for tailoring the selection of information to be recorded and reported on item configuration status. (MIL-STD-482.)

5. Uniform terminology and definitions for configuration management. (MIL-STD-480.)

6. Proposed Air Force standards (which supersede AFSCM 375-1) to be used with the preceding requirements are presently under CODSIA review. (MIL-STD-483.)

Impact on Existing Governing Documents

The new standards impact to varying degrees on a wide variety of configuration management policies, practices, and procedures now in use; for example, at least 23 DOD component documents have been identified as providing configuration management guidance to their activities for internal use. Some of these documents may be consolidated or eliminated; the remainder will require some revision to reflect the improvements intended.

For contractual application approximately 70 DOD component documents have been identified for use. Of these documents, 29 will no longer be authorized for use on new contacts; the remainder will be revised to be compatible with the new policies and practices.

Equally important, a basis has now been established to assure that any existing or new configuration management documents to be applied on contracts will be authorized by the Office of the Secretary of Defense before use.

To summarize the above, given these new standards we may obviously expect increased uniformity among the services in terminology and methodology.

22.6   HAVE WE DISENGAGED?

Two areas that attract one's attention in examining the new DOD standards are as follows: a relaxation of detail instructions for standard hardware and data software identifiers such as originally espoused by Exhibits X and XII of AFSCM 375-1[5]; and the optional use of the allocated baseline (previously known as the design requirements baseline).

The optional use of the allocated baseline for the allocation of a complex item's functional characteristics to its major components offers interesting possibilities, however speculatory.

Let us assume an aircraft system for purposes of our discussion. Level 1 is the aircraft system establishing the functional baseline (Figure 22.2). The air vehicle is the allocated baseline, as are other level 2 items. Air frame,

[5] Exhibit X, AFSCM 375-1, "Standard Configuration Identification Numbers." Exhibit XII, AFSCM 375-1, "Engineering Release Record Requirements."

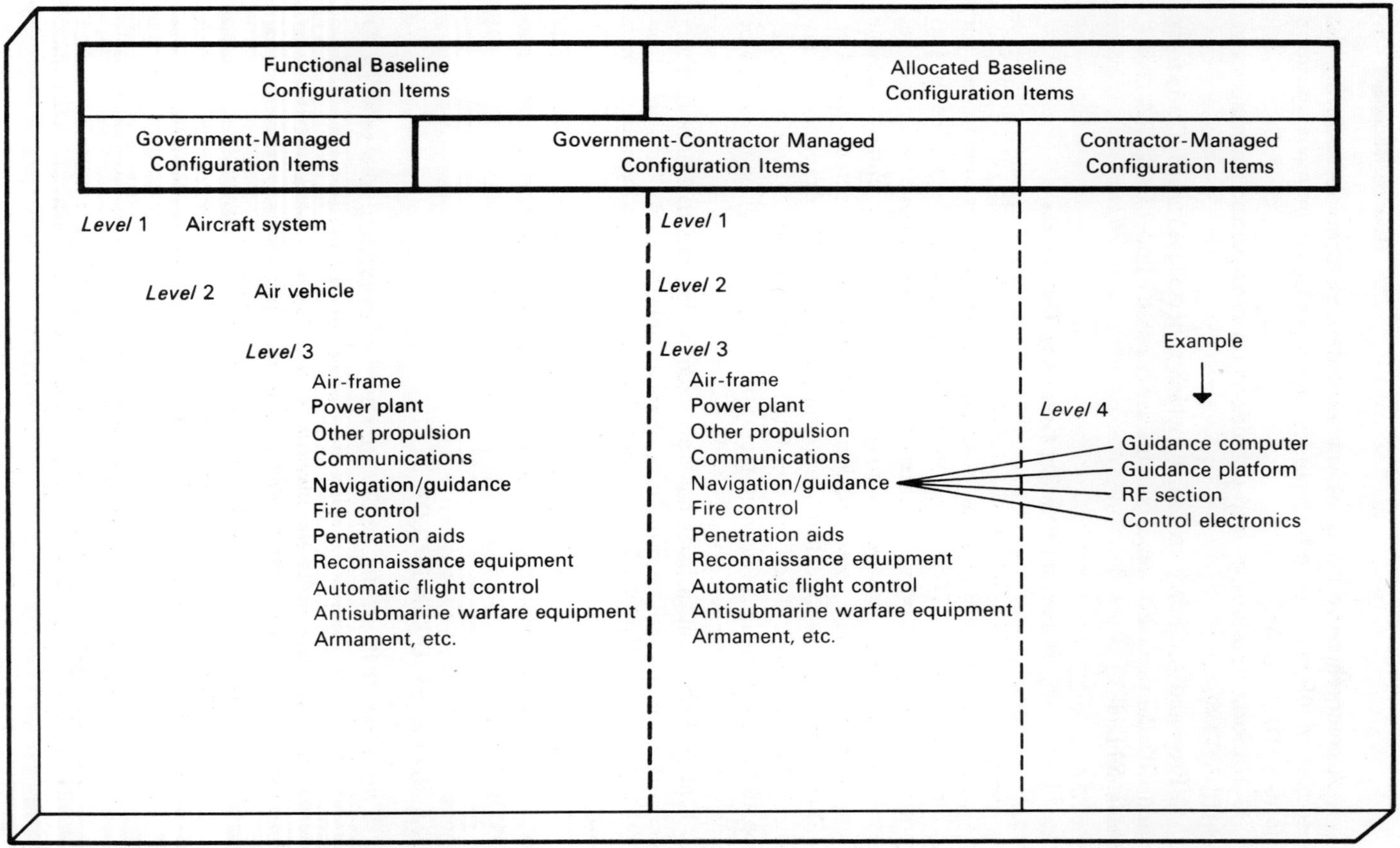

Functional Baseline Configuration Items

Government-Managed Configuration Items	Government-Contractor Managed Configuration Items

Allocated Baseline Configuration Items

Contractor-Managed Configuration Items

Level 1   Aircraft system

Level 2   Air vehicle

Level 3

Air-frame
Power plant
Other propulsion
Communications
Navigation/guidance
Fire control
Penetration aids
Reconnaissance equipment
Automatic flight control
Antisubmarine warfare equipment
Armament, etc.

Level 1

Level 2

Level 3

Air-frame
Power plant
Other propulsion
Communications
Navigation/guidance
Fire control
Penetration aids
Reconnaissance equipment
Automatic flight control
Antisubmarine warfare equipment
Armament, etc.

Level 4

Example →

Guidance computer
Guidance platform
RF section
Control electronics

Figure 22.3 Concept of hierarchical CI management levels.

propulsion, and so forth, are level 3 work breakdown structure items,[6] also to an allocated baseline. As established in this text, the systems engineering effort during the definition phase would have resulted in Part I specifications for all the major configuration items. These would now be called allocated baselines under the new terminology and would have normally been generated for items in levels 2 and 3. Further, these Part I specifications would be approved by the government, made part of the engineering development contract, and subject to formal ECP control. Or, to put it another way, both the contractor and the government would elect to manage items in the second and third indenture levels under the old system.

Under the new concept the process of the definition phase may result in functional allocations to equipment levels not envisioned at the time of the functional baseline, hence the optional allocation baseline. Mutual contractor and government agreement to manage at the new (lower) equipment level then brings us back to the old system.

But most intriguing is the possibility that under the new concept of baseline application we may have the opportunity to differentiate between government-managed items and contractor-managed items (although nowhere is this specifically implied in the new standards). To carry our speculation further, the government will manage the allocated baseline only at the level at which they are contracting—the level established at the functional baseline. It is then encumbent on the contractor to prove the validity of any allocated baseline he elects to establish. It is *his* baseline, not the government's. He does not seek, nor is he required to seek, government approval of changes. The contractor controls the tasks; for example, the contractor-managed CI's within the government-contractor managed navigation/guidance sybsystem (see Figure 22.2) may be the guidance computer, guidance platform, RF section, and control electronics. One by-product—and an important one to the contractor—is the ability to make decisions and trade-offs without government intervention. Perhaps for the first time, incentive contracting theory and practice will be in harmony with configuration management. While we are speculating, however, it is important to note that even though the government perhaps may not be a party to the allocated baseline during the development effort, the government will manage the *product baseline* at the *allocated* CI levels, including spares.

As was discussed earlier, the ideal degree of government engagement in management controls, such as configuration management, can be achieved by what we called "responsive visibility." Given that our speculation of differential hardware management levels between the government and

---

[6] MIL-STD-881, "Work Breakdown Structures." DOD Directive 5010.20, "Work Breakdown Structures for Defense Material Items."

contractor were true—poor management on the part of the contractor would prompt the government to go deeper into the indenture levels of a system during the development effort and redefine the allocated baseline. This, we feel, suggests the single major impact of the contemporary scene on configuration management as it has been derived since June 1962.

Chapter 23

EPILOG

Chapter 22 completed the technical aspects of this book; we have covered how we started, where we are, and where we are going in the discipline of configuration management. We have tried to present a balanced mixture of generalized concepts and detailed procedures in order to meet the needs of both those interested in the theoretical aspects of the discipline as well as those interested in the specifics. We have covered a broad range of topics and necessarily have not gone into the details of each. The most important concepts, we think, are progressive baseline management, formalized change control of source documents, and identification of all products and related paper work. We have made practical judgments based on our experiences as to which topics were to be treated in detail and which were not. Other writers in this discipline may have taken another approach with different topics emphasized based on their own experiences and special interests. However, the subject matter given represents a comprehensive coverage of the contemporary configuration management scene and prepares engineers and managers for intelligent participation in configuration identification, control, and accounting activities.

Remember that configuration management is primarily an art, not a science, and therefore the rules and methods described in the previous chapters are not sacrosanct; they can be changed when analyzed and weighed in terms of your needs. Of course, when they are written into a contract they become sacred by necessity. In either case, the exact system and methods laid down for configuration management are not as important as having a system that works and is not in constant flux, with the confusion that will inevitably result. Random or lightly thought-out experiments with the system can be dangerous because its effective application rests with a project team that knows its rules and can thus implement them without constantly relearning new methods while trying to produce an equipment or CI under a tight schedule and heavy work load.

Although the orientation of this book is primarily towards the aerospace industry, we want to reiterate that the information and developments presented have a broader application for progressive managers and engineers in the commercial world. The topics discussed are in general not new but represent what has been going on in good business practice for years, though in a less structured and defined way. What is new is the union and consolidation of these practices with the newly developed methods used by aerospace companies building sophisticated products to form a new discipline called configuration management. This discipline, although referred to as "esoteric" in the first chapter, is one that cuts across the entire spectrum of engineering, production, and management activities. Whether you have a configuration management department or not, you must have a grasp of the configuration management system and procedures described to perform your job effectively when your responsibilities span more than one area of your company's activities.

This discipline is the culmination of years of specialized experience and millions of dollars spent by the people of the United States over the last 30 years. Besides the spectacular achievements that we have made in space and related fields, it would be of great benefit to us if American industry could selectively apply the information presented in the book to improve the products that are supplied to the consumer—whether these products are pencils, shoes, automobiles, irons, or homes. Although not a cure-all, the major point of the discipline is to help produce better products more efficiently within the framework of competition, acceptable cost, and the needs of the customer, the American people.

BIBLIOGRAPHY

PART I GENERAL

Affel, H. A., Jr., "Systems Engineering," *International Science and Technology*, November 1964, pp 18–26.

American Society for Quality Control, *"West Coast Configuration Management Symposium Technical Papers,"* August 19–20, 1965.

Archibald, R. D., and R. L. Villoria, *Network-based Management Systems (PERT/CPM)*, Wiley, New York, 1967.

Bashaw, C. J., and C. E. Gardella, "AFSCM 375-1, In Retrospect," ESD-TR-67-230, March 1967, Technical Requirements and Standards Office, Electronic Systems Division, AFSC.

Baumgartner, J. S., *Project Management*, Irwin, Homewood, Ill., 1963.

Bonini, C. P., *Management Controls: New Directions in Basic Research*, McGraw-Hill, New York, 1963.

Budd, A. E., "A Method for the Evaluation of Software: Executive, Operating or Monitor Systems," *The Mitre Corporation*, September 1967, AD 661 999, distributed by Clearinghouse for Federal Scientific and Technical Information.

Bunker, W. B., Lt. General, USA, "Objectives of Configuration Management," *Defense Industry Bulletin*, September 1967, pp 1–3.

Canning, R. G., Ed., "Overall Guidance of Data Processing," *EDP Analyzer*, August 1968, Vol. 6, No. 8.

Canning, R. G., Ed., "Managing the Programming Effort," *EDP Analyzer*, June 1968, Vol. 6, No. 6.

Cleland, D. I., and W. R. King, *Systems Analysis and Project Management*, McGraw-Hill, New York, 1968.

Cleland, D. I., "Why Project Management," *Business Horizons*, Winter 1964.

Dablin, R. C., "A Corporate Part Numbering System," *Standards Engineering*, December 1960–January 1961, pp 3–7.

Engoron, E. J., and A. L. Jackson, Jr., "Uniform Policy and Guidance Established for Configuration Management," *Defense Industry Bulletin*, January 1969, pp 1–41.

Geddes, P., "Customer Closes Loopholes In Program Management," *Aerospace Management*, April 1964, pp 50–54.

Greene, *Production Control: System and Decisions*, Irwin, Homewood, Ill., 1963.

Gruenberger, F., "Program Testing and Validating," *Datamation*, July 1968, pp 39–47.

Hantz, E. C., and A. E. Lager, "Configuration Management—Its Role in the Aerospace Industry," RATA Article R69-14223, 1968 Product Assurance Conference.

Hitch, C. J., "An Appreciation of Systems Analysis," RAND Corporation, DDC AD422837, Santa Monica, Calif., August 18, 1955.

LaBolle, V., "Management Aspects of Computer Programming for Command and Control Systems," *System Development Corporation*, February 5, 1963, SP-1000/000/02.

Laine, M. J., and E. C. Spevak, "Configuration Management," *Space/Aeronautics*, November 1966, pp 74–81.

Lanier, F., "Organizing for Large Engineering Projects," *Machine Design*, December 27, 1956, p 54.

Liebowitz, B. H., "The Technical Specification—Key to Management Control of Computer Programming," *Spring Joint Computer Conference*, 1967, pp 51–59.

Mazia, J., and J. V. Panek, "Numbering Systems," Annual Meeting of the Standards Engineering Society, Pittsburgh, 1960.

Nucci, E. J., and A. L. Jackson, Jr., "Work Breakdown Structures for Defense Materiel Items," *Defense Industry Bulletin*, February 1969, pp 22–28.

O'Donnell, C. J., *Principles of Management*, 3rd ed., McGraw-Hill, New York, 1964.

Piligian, M. S., and J. L. Pokorney, "Air Force Concepts for the Technical Control and Design Verification of Computer Programs," *Spring Joint Computer Conference*, 1967, pp 61–66.

Ratynski, M. V., "The Air Force Computer Program Acquisition Concept," *Spring Joint Computer Conference*, 1967, pp 33–44.

Riley, David E., Brig. General, USAF, "The Government's Role in Minding Its Contractor's Business," *Defense Industry Bulletin*, April 1968, pp 20–22.

Roberts, E. B., "How the U.S. Buys Research," *International Science and Technology*, September 1964, pp 70–77.

Schoderbek, P. P., *Management Systems*, Wiley, New York, 1967.

Searle, L. V., and George Neil, "Configuration Management of Computer Programs by the Air Force: Principles and Documentation," *Spring Joint Computer Conference*, 1967, pp 45–49.

Seith, W., Captain, USN, "Configuration Management in the Navy," *Defense Industry Bulletin*, April 1967, pp 4–7.

Tupac, J. D., "An Approach to Software Evaluation," *American Management Association Annual EDP Conference, New York*, March 6, 1967.

PART II   GOVERNMENT REQUIREMENTS DOCUMENTS

AF Regulation 57-4	"Modification/Modernization of Systems and Equipment"
AF Regulation 65-3	"Configuration Management"
AFSC (SCS-22) Letter, 25 February 1969	"Interim Policy Guidance on Configuration Management"
AFSCM 375-1 (Manual), Air Force	"Configuration Management During Definition and Acquisition Phases"
AFSCM 375-4 (Manual)	"Systems Program Management"
AFSCM 375-5 (Manual)	"Systems Engineering Management Procedures"
AFSCM/AFLCM 310-1, (Manual)	"Data Management"
AFSCP 70-1	"Lessons Learned from AF Contractor Procurement System Review"
AFSCR 57-2/AFLCR 57-24	"Operational Requirements—Configuration Management During Acquisition Phase"
AMCR 11-26 (Manual), Army	"Configuration Management"
AMCR 715-29	"Control of Engineering Changes Under Competitively Obtained Fixed-Price Supply Contracting"
ANA Bulletin 445	"Engineering Changes to Weapons, Systems, Equipments, and Facilities"

BUSHIPS Document DS & FM Manual
    4760.3 — "Ship Building and Conversion Contract Modification"

BUWEPSINST 4121.2 — "Specifications Under the Cognizance of the Bureau of Naval Weapons"

BUWEPSINST 4340.2 — "Aircraft Master Configuration Lists"

BUWEPSINST 4440.9 — "Backfitting and Modernization of Weapon System Material"

BUSHIPSINST 4720.1 — "Alterations to Naval Vessels"

BUSHIPSINST 4760.21 — "Issuance of Changes Affecting Ship Contractors and Conversion Funds"

BUWEPSINST 4760.25 — "Shipbuilding Materiel Management Systems"

BUWEPSINST 5200.10A CP-3 Navy — "Weapon System Changes; Policy, Origination, Responsibilities and Procedures for"

BUWEPSINST 5200.14 — "Surface Missile Systems Engineering Change Procedures"

BUWEPSINST 5200.20 Navy — "Weapon System Configuration Control Manual"

BUWEPS-SMS-INST 5200.6 — "Surface Missile Systems Class I and Class II Design Change Control"

CM2A — "Configuration Management," US Army Missile

DOD Directive 4100.35 — "Integrated Logistics System"

DOD Directive 5010.14 — "System/Project Management"

DOD Directive 5010.19 — "Configuration Management"

DOD Directive 5010.20 — "Work Breakdown Structures for Defense Material Items"

DOD Instruction 5010.21 — "Configuration Management"

M-200B, Chapter V — "Defense Standardization Manual (Specifications)"

MIL-D-1000 — "Drawings, Engineering, and Associate Lists"

MIL-M-9868 — "Microfilming of Engineering Documents, 35mm, Requirements for"

MIL-S-83490 — "Specification Acquisition Practices"

MIL-STD-12 — "Abbreviations for Use On Drawings and in Technical Type Publications"

MIL-STD-100 — "Engineering Drawing Practices"

MIL-STD-129 — "Marking for Shipment and Storage"

MIL-STD-130 — "Identification Marking of US Military Property"

MIL-STD-280 — "Definition of Equipment Divisions"

MIL-STD-480 — "Engineering Changes, Deviations, Waivers"

MIL-STD-481 — "Engineering Changes (Short Form)"

MIL-STD-482 — "Configuration Status Accounting Data Elements and Related Features"

MIL-STD-490 — "Specification Practices"

MIL-STD-499 — "Systems Engineering"

MIL-STD-881 — "Work Breakdown Structure for Defense Material Items"

NAVMATINST 4000.15 Navy — "Management of Technical Data and Information"

NAVMATINST 4130.1 Navy   "Configuration Management"

NAVORD OSTD 65   "Ordinance Alterations and Modifications, Preparation and Issuance of"

NPC-500-1 (Manual), NASA   "Apollo Configuration Management Manual"
(Now NHB 8040.2)

ACRONYMS

Acronyms commonly used in configuration management and related fields are listed below. Because each major company and Government agency uses its own acronyms, it is impossible to provide a complete list. Remember that companies may assign different meanings to the same acronyms. Government acronyms, however, are generally more consistent in meaning.

(A)	alphabetic designator
AAE	aerospace ancillary equipment
ABM	advance bill of materials
A & T	assembly and test
AC	associate contractor
ACE	acceptance checkout equipment
ACI	allocated configuration identification
ACO	administrative contracting officer
ADCN	advance design change notice
ADO	advanced development objective
ADP	acceptance data package or automatic data processing
ADPE	automatic data processing equipment
AF	Air Force
AFR	Air Force regulation
AFLC	Air Force Logistics Command
AFM	Air Force Manual
AFPR	Air Force plant representative
AFSC	Air Force Systems Command
AFSCM	Air Force Systems Command Manual
AFSSC	Armed Forces Supply Support Command
AGE	aerospace ground equipment
AIA	Aircraft or Aerospace Industries Association (see AIAA)
AIAA	Aerospace Industries Association of America, Inc.
AIEE	American Institute of Electrical Engineers
AIR	aerospace information report
ALM	advance list of materials

ALT	alteration
AMC	Army Materiel Command of Air Materiel Command
AMCM	Air Materiel Command Manual
AMCR	Air Materiel Command regulation
AMS	aerospace material specification
AN	Air Force-Navy or Army-Navy
ANA	Air Force-Navy-aeronautical or Army-Navy-aeronautical
(A/N)	alpha/numeric designator
AND	Air Force-Navy design or Army-Navy design
AQL	acceptable quality level
AR	as required
A/R	as required
AR	Army regulation
ARC	Ames Research Center (NASA)
ARP	aerospace recommended practice
ARS	American Rocket Society
AS	aerospace standard or aeronautical standard
ASA	American Standards Association (see USASI)
ASG	Aeronautical Standards Group
ASPR	armed services procurement regulation
ASQC	American Society of Quality Control
ASTIA	Armed Services Technical Information Agency
ASTM	American Society for Testing Materials
ATP	acceptance test procedure or approval to proceed
AVE	aerospace vehicle equipment
AWS	American Welding Society
BOB	Bureau of Budgets (now Office of Management and Budget)
BOD	beneficial occupancy date
BOI	break of inspection
CAI	configuration acceptance inspection
CAR	configuration audit review or corrective action request
CAS	contract action status
CCB	configuration control board or change control board
CCBD	configuration control board directive
CCCB	configuration change control board
CCN	contract change notification or notice
CCO	contract change order
CD	classification of defects
CDR	critical design review or conceptual design review
CDRL	contractor data requirements list
CECB	change evaluation control board
CEI	contract end item
CEIN	contract end item number
CFE	customer or contractor furnished equipment
CIN	change or configuration identification number
CI	configuration item (same as CEI) or critical item
CII	configuration item identification or configuration identification index
CIP	cost improvement proposal

CIPL	CI parts list
CIR	change incorporation record
CMD	configuration management directive or division
CMM	configuration management manager
CMO	configuration management office or contracts management office
CMR	configuration management region
CO	contracting officer
CODSIA	Council of Defense and Space Industries Association
CPB	change planning board (same as CPG)
CPC	computer program component
CPCI	computer program configuration item
CPG	change planning group (same as CPB)
CPI	computer program identification or change package identification
CR	change request
CRN	contract release notice (sales order) or change request number
CRS	calibration requirements summary
CSAR	configuration status accounting report
CVBD	configuration verification baseline determination
DCAA	defense contract audit agency
DCAS	defense contract administration services
DCASR	defense contract administration services region
DCCR	design change completed report
DCN	design change notice
DDC	defense documentation center
DESC	defense electronics supply center
DI	data item
DIAC	Defense Industry Advisory Committee
DL	data list
DOD	Department of Defense
DODD	Department of Defense directive
DODISS	DOD Index of Specifications and Standards
DREO	design revision engineering order
DRL	data requirements list
DRN	documentation revision notice
DRM	drafting room manual
DS-RPIE	direct support real property installed equipment
DTA	development test article
DVU	design verification unit
EAPL	engineering assembly parts list
ECA	engineering change analysis
ECI	engineering change instruction (same as EO)
ECN	engineering change notice
ECO	engineering change order
ECP	engineering change proposal
ECR	engineering change request
ECRA	engineering change request and analysis
EIA	Electronic Industries Association
EID	end item designator or end item description
EM	engineering model

EO	engineering order
EPOE	end piece of equipment
ER	engineering request
ETR	Eastern Test Range
FAC	facility or first article configuration
FACI	first article configuration inspection
FACR	first article configuration review
FAD	first article demonstration
FAI	first article inspection
FAIA	first article inspection acceptance
FCEI	facility contract end item
FCA	functional configuration audit
FCI	functional configuration identification
FCR	final configuration review or facility change request
FDN	family designation number
FDR	final design review
FED-SPEC	federal specification
FED-STD	federal standard
FIIN	federal item identification number
FL	flight model
FLT	flight model
FMECA	failure mode, effect, and criticality analysis
FR	failure report
FRR	flight readiness review
FSC	federal supply classification, supply code, or stock class
FSCM	federal supply code for manufacturers
FSE	factory support equipment
FSN	federal stock number
FY	fiscal year
G & A	general and administrative
GIE	ground instrument equipment
GFAE	government furnished aerospace equipment
GFE	government furnished equipment
GFP	government furnished property (connectors, cables, CI's and so on)
GQI	government quality inspector
GQR	government quality representative
GSE	ground support equipment
GSFC	Goddard Space Flight Center (NASA)
IAC	integration, assembly, and checkout or integrating associate contractor
I & C	installation and checkout
I & M	installation and maintenance
I & S	interchangeability and substitution
ICD	interface control drawing or document
ID	deck information
IDL	indentured drawing list
IL	index list
ILS	integrated logistics support

I/O	input-output
IPB	illustrated parts breakdown
IR	inspection report
IRN	interface revision notice
ITO	interim technical order
ITP	in-process test procedure
JAN	joint Army-Navy
JANAF	joint Army-Navy-Air Force
JPL	Jet Propulsion Laboratory (NASA)
KSC	John F. Kennedy Space Center
KSN	kit stock number
LCP	letter change proposal
LM	list of materials (same as parts list)
L/M	list of materials
LRC	Langley Research Center (NASA)
LS	life support
LVL	level of organization to accomplish retrofit; for example, factory, depot, field, or operations
MCP	manufacturing change point or military construction program
MCMSP	military communication manual specification publications
MDS	mission, design, series
MFG	manufacturer
MGE	maintenance ground equipment
MH's	man hours
M/H	man-hours
MIL	military
MIL-HDBK	military handbook
MIL-SPEC	military specification
MIL-STD	military standard
MIP	material improvement project
MIPR	material interdepartmental procurement request
MIRR	material inspection and receiving report (DD Form 250)
MO	manufacturing order
MOD	modification or model
MPD	missile purchase description
MRB	modification or material review board
MRR	multiple reference release
MS	military sheet standard
MSC	Manned Spacecraft Center (NASA)
MSE	maintenance support equipment
MSFC	Marshall Space Flight Center (NASA)
MSS	military supply standard
MTBF	mean time between failures
MWO	modification work order
(N)	numeric designator
NA	not applicable, next assembly, or not available

N/A	not applicable
NAS	National Aerospace Standards or National Aircraft Standards
NASC	National Aircraft Standards Committee
NASA	National Aeronautics and Space Administration
NAVAIR	Naval Air Systems Command
NAV MAT INST	naval materiel instruction
NAVORD	naval ordnance
NAVSHIPS	naval ships
NAVWEPS	naval weapons
NC	no change
N/C	no change
NCR	nonconformance report
NEMA	National Electrical Manufacturers Association
NHA	next higher assembly
NHB	NASA handbook
NOR	notice of revision
NPC	NASA publication control number
NR	not required
N/R	not required
NHQ	NASA Headquarters
ODC	other direct costs
OGE	operating ground equipment
O & M	operation and maintenance
OPS	operational
OSE	operational support equipment
OSR	operational support requirement
P	prototype
PBS	program breakdown structure
PC	prime contractor
PCA	physical configuration audit (as-built check)
PCD	parts control drawing (same as SCD)
PCI	product configuration identification
PCN	procedure change notice
PCO	procedure change order, procuring contracting office, or procurement contracting officer
PCP	program change proposal
PCR	product conformance review
PDP	preliminary definition plan
PDR	preliminary design review
PERT	program evaluation and review technique
PI	principal investigator
PL	parts list
P/L	parts list
PMI	program management instruction
PN	part number
P/N	part number
PO	purchase order
PPB	preliminary parts breakdown or provisioning parts breakdown

PR	purchase request
PTDP	preliminary technical development plan
PTM	prototype model
PTP	production test procedure (same as ITP)
QA	quality assurance
QPL	qualified products list
QSL	qualification status list
QTP	qualification test procedure
R & D	research and development
RDT & E	research, development, test, and evaluation
RAS	requirements allocation sheet
RFI	radio frequency interference (same as EMI, electromagnetic interference)
RFP	request for proposal
RFQ	request for quote
RN	revision notice
RPIE	real property installed equipment
RTA	request for technical action (similar to an ECP)
RTV	return to vendor
SAR	special aeronautical requirements
SAE	Society of Automotive Engineers
SC	service change (similar to a field or retrofit change); also spacecraft
S/C	spacecraft
SCD	specification or source control drawing
SCDN	specification or source control drawing number
SCN	specification change notice
SCNP	specification change notice proposal
SCCN	subcontract change notice
SCR	software change request
SDD	system definition directive
SDRL	subcontractor data requirements list
SE	support equipment
S/E	system or equipment
SFA	system functional audit
SN	serial number
S/N	serial number
SOR	specific operational requirement
SOW	statement of work
SP	spares
SPL	system parts list
SPCM	software program configuration manager
SPD	system program directive or director
SPO	system program office
SSM	system support manager
STAR	Scientific and Technical Aerospace Reports (NASA)
STD	standard
STE	special test equipment
SV	space vehicle

SVE	space vehicle equipment
TA	top assembly or type approval model
T & M	time and material
TAR	technical action request (similar to ECP)
TATP	type approval test procedure (same as QTP)
TBD	to be determined
TCPI	tactical computer program identification
TCR	test change request or record
TCTO	time compliance technical order
TCTR	time compliance technical requirement
TD	technical data or document
TDP	technical data package (similar to ADP) or technical development plan
TDR	technical documentation report
TM	technical manual
TMS	type, model, series
TO	technical order
TRB	test review board
TRE	training equipment
TRS	technical requirements specification
TWX	teletypewriter exchange
UR	unsatisfactory report
USASI	United States of America Standards Institute (formerly ASA)
USP	uniform specification program
VCR	verified configuration record (as-built list)
VE	value engineering
WBS	work breakdown structure
W/P	work package
WR	weapons requirements
WSMR	White Sands Missile Range
WTR	Western Test Range

GLOSSARY

This glossary contains definitions of common configuration management terms. A list of key acronyms precedes the glossary.

Acceptance, the process by which the customer's representative formally agrees to ownership of a completed equipment.

Acceptance data package describes the form, fit, manufacturing requirements, characteristics, and performance history of an equipment. The package includes specifications, drawings, plans, manufacturing orders, inspection data, test procedures, test data, and other data required by the customer.

Acceptance test procedure, used to verify that the equipment meets contract requirements before it is accepted and delivered to the customer.

Advanced development refers to projects that have proceeded into the development of hardware for experimental or operational tests.

Acquisition phase, the period between the end of the definition phase and the delivery of the last equipment to the customer.

Advance design change notice, a document describing an approved drawing revision that has not been incorporated onto the drawing.

Advance release drawing allows procurement of long leadtime items to begin when the design is only partially complete. It helps avoid schedule slippage if procurement is delayed until a production drawing is released.

Allocated baseline, an allocated configuration identification which is an optional baseline initially approved by the customer. See allocated configuration identification.

Allocated configuration identification, performance specifications guiding the development of configuration items that are a part of a higher level CI. These specifications cover functional characteristics allocated from those of the higher level CI, tests to demonstrate achievement of the functional characteristics, interface requirements, and design constraints.

Allocation document defines actual or planned geographic location of each product built during the project.

Alphanumeric identifier, a combination of letters and numbers that identifies a product, item, or document.

Ancillary document, a design description document, made after a drawing is formally released, which changes the design description of the released drawing and is identified and released as a part of the drawing; e.g., EO, ADCN, or DCN.

Annotated listing, a listing describing the computer program in terms of its detail elements: decks, sequences, storage requirements, and so on.

Article, a product, equipment, or item.

Assembly, a unit of two or more related subassemblies designed to perform a specific function and capable of disassembly.

Assemblying, the accumulation of main and auxiliary memory portions of a computer incoming long message.

Associate contractor, a contractor responsible for a CI in a system and, reporting to the system program director.

Associated lists, parts list, wire list, data list, and index list.

Audit, to inspect records and procedures.

Baseline, an approved reference point for control of future changes to a product's performance, construction, and design. Mainly specifications and drawings.

Batch, the same as lot.

Black box, a product element that identifies a functional requirement but does not reveal anything about its contents or construction.

Bonded storeroom, a controlled storeroom for CI parts and materials.

Breadboard, a primitive model of an equipment used to demonstrate that the design meets performance requirements. Breadboard packaging and construction do not conform to the flight configuration and workmanship standards.

Break of inspection, a record that describes all replacement or rework done to an equipment after it has been officially inspected and approved for completeness and correct configuration.

Buyer, the same as a customer.

Byte, any group of binary digits that can be separately manipulated within a binary (computer) word.

Calling sequence, the sequencing and ordering of items in computer file records for examination on a selected basis.

CEI or CI number, a permanent number assigned to a product for identification.

Change control board, the same as configuration control board.

Change documentation, specification change notice, engineering order, engineering change proposal, design change notice, notice of revision, and so on.

Change identification number, a number assigned to a data package defining an equipment engineering change. It is used to control, sequence, and account for production, implementation, and retrofit actions related to the change. The CIN includes the CI number, company code identification number, ECP number, ECP type code, ECP revision code, and ECP correction code.

Change request (engineering change request), a document that is used by the project staff to request a change to the approved product configuration.

Check sum, an arithmetic summary of memory cells in a bit format, thus representing the authorized configuration of the computer program.

Checkout, the process for determining whether an equipment is capable of performing its required functions.

Class I change, a change affecting the contract specification, price, weight, delivery schedule, reliability, performance, interchangeability, interface with other products, safety, RFI, or GSE.

Class II change, any change not falling within the Class I change definition given above.

Class 1 drawing, a drawing for which the government retains responsibility for preparation and maintenance.

Class 2 drawing, a drawing for which the company retains responsibility for preparation and maintenance.

Classification of defects, refers to critical, major, or minor deficiencies in the equipment.

Code identification, a unique 5-digit government identification number issued to each company that builds or develops items for the government.

Cognizant engineer, the engineer responsible for a particular technical area such as electronic design, testing, or packaging.

Commercial item, an item regularly used for other than government purposes and sold during normal business operations.

Company standard, a specification prepared by the company and used for many projects or applications. See standard.

Component, a part, subassembly, assembly, or combination of these items joined together to perform a function.

Compatibility ECP, an ECP priority used for changes required during system installation and checkout that are necessary to make the system work (design deficiency correction). Also used to process changes to system requirements after the design requirements baseline is established (not a design deficiency).

Concept formulation, the effort made before a decision to conduct engineering development. Includes system studies and experimental hardware tests.

Conceptual phase, the period preceding the definition phase. It begins with determination of broad product objectives and ends with the start of definition phase; it includes concept formulation, general equipment design approach, feasibility evaluation, block diagram, and equipment layout. This phase is usually conducted by the customer before the equipment contract is released to a company.

Configuration, the complete technical description required to build, test, accept, operate, maintain, and logistically support an equipment. Also the physical and functional characteristics of the equipment.

Configuration accounting, the reporting and recording of changes made to the approved configuration.

Configuration audit review, a technical review comparing each CI documentation description with the prototype to assure the documentation's accuracy and adequacy for manufacture and its conformance to the CI description prepared during the development effort. Similar to FACI.

Configuration control, the evaluation, coordination, and approval of all changes to the equipment configuration defined by the baseline.

Configuration control board, a group of technical and administrative project personnel who are responsible for reviewing and assessing engineering changes to the CI after the baseline has been approved.

Configuration element, an item subject to configuration management.

Configuration element identifier, the alphanumeric designator assigned to identify a configuration element; for example, the identifiers for an assembly, a subassembly, or any other item that is subject to configuration control.

Configuration identification, the technical data describing the approved configuration of the product or the process for identifying these data, the product, and changes made to them.

Configuration identifier, an alphanumeric designator used to identify configuration elements; the same thing as a configuration element identifier.

Configuration item, a collection of hardware or software, or any of its parts, that satisfies an end use and is designated by the government or customer for configuration management; the same as CEI and CI.

Configuration management reports, a collection of documents that describe the status of the equipment configuration.

Configuration management, the art of providing systematic and uniform configuration identification, control, and accounting of an equipment and its parts.

Configuration management directive, a document that records the decision of the CCB and is the vehicle for official ECP approval or rejection by the customer.

Configuration management office, the organization within the system program office that is responsible for (a) formulating, issuing, and maintaining all configuration management documentation; (b) administration support to the CCB; (c) direction and supervision of the uniform specification program; and (d) the transferring of all configuration documentation to the customer. A CMO is not generally required for a small project but is necessary for a large program.

Configuration number, the same thing as a configuration identifier.

Configuration status and accounting index, a record of all engineering changes (EO's or ECP's) to an equipment; it may also include cancelled and rejected ECP's and EO's.

Configuration status, the actual configuration of an equipment at a given time in relation to an approved configuration or baseline.

Contract, a legal document that establishes agreement between the customer and company defining the equipment to be built, other tasks to be done, and the terms and conditions under which the work will be performed.

Contract administrator, an individual authorized to enter into and administer the contract.

Contract change notice, a written order signed by the customer's contract administrator directing the company to proceed with changes requested by the company.

Contract definition, normally a competitive period or phase which involves the verification or completion of preliminary design of a CI. Includes firm contract and management planning.

Contract end item, a deliverable item that is formally accepted by the customer. The same as CI.

Contract maintenance, the procedure for changing a contract to incorporate approved changes and corrections to the text, including preparing documents for recording the authorization and history of changes and revisions.

Contract (procurement) specification, a customer-prepared document that defines the characteristics, performance, and operating limits of a CI or equipment.

Contractor, the company designing, developing, and building the equipment for the customer or government. Same as seller.

Contracts office, the office responsible for all legal and financial aspects of the project.

Controlled drawing, a drawing that has been through the authorized preparation, review, approval, and release cycle and whose distribution and revision are controlled by the data control group or department.

Core budget, the predicted amount of storage available for information destined for the computer memory (core storage).

Cost refers to the dollars paid by the government for an item or service.

Cost analysis, an itemized list of labor, materials, test equipment, tooling, fixtures, and documentation required to make an engineering change to an equipment.

Cost effectiveness, the property of a design or administrative change that (a) will improve the capability of a product at minimum increased cost or (b) will reduce product capability by an acceptable amount with a resulting major cost reduction. Also, the measure of an item's ability to fulfill a specific need at minimum cost.

Critical component, a component within a CI that requires an approved specification to establish technical or inventory control at the component level.

Critical design review, a formal technical review of design that is done to identify specific engineering documents for release to production and to establish a basis for spares provisioning, preparation of manuals, and other support activities depending on the detail definition of the equipment.

Critical deviation, a departure from documentation affecting safety of the CI. (See deviation.)

Critical waiver, a departure from documentation affecting the safety of the CI. (See waiver.)

Criticality, the relative importance of items within an equipment to its successful operation.

Data, the media for communicating ideas, descriptions, requirements, plans, and instructions related to a project, hardware, or software. Data items include drawings, parts and wire lists, manuals, specifications, standards, reports, punched cards, and computer print-outs.

Data chain, a grouping of two or more data elements used for configuration status accounting reports. For example, a type, model, series designator consists of three data elements: type, model, and series.

Data code, a letter, number, symbol, or any combination of these elements used to represent a data item, such as P for production, R for retrofit, and K for contract definition.

Data element, a grouping of informational items used for configuration status accounting reports; for example, part number, serial number, quantity, series designator, time, man-hours, and modification kit number.

Data item, a subunit of descriptive information classified under a data element in configuration status reports. These include technical orders, specifications, standards, requests for proposals, reports, and engineering development.

Data use identifier identifies the use of a data element in a configuration status report. For example, a supplementary procurement identification number could have the following data use identifiers: (a) amendment numbers for solicitation document and (b) modification numbers for control change agreement.

Debugging, the examining or testing of a procedure, program, or equipment for detecting and correcting errors.

Deficiencies refer to (a) the characteristics of any item that do not comply with the specified configuration or (b) inadequate configuration data that may cause a CI to not meet its operational requirements.

Definition baseline, a competitive period between conceptual and acquisition phases (or periods) which results in completion of Part I of the detail product specification (firm performance requirements).

Design activity, the group responsible for the design of an item or equipment.

Design and development phase, the period covering design, development, and testing of engineering and prototype hardware during the acquisition phase. Begins during the definition phase and normally ends upon completion of FACI.

Design calculations are calculations used to demonstrate that product design characteristics or goals will be achieved when the equipment is built.

Design change notice, a record of revisions made to a drawing original.

Design change request, the same thing as a change request.

Design constraints, envelope dimensions, weight, shape, mounting configuration, RFI characteristics, component standardization, use of inventory items, and integrated logistics support policies to be followed.

Design criteria, specific standards or goals for design of the product; for example, lightweight, compactness, high reliability, simplicity, safety, low power consumption, high accuracy, long operating life, ease of maintenance, flexibility, and versatility.

Design inventory, written analyses of new or revised designs current at the time of a design review.

Design requirements baseline, a baseline for a CI that is technically defined by Part I of the detail equipment specification.

Design requirements ECP proposes a change to Part I of equipment detail specification after it has been approved to define the equipment design requirements baseline.

Design review, an examination of the equipment design, construction, data, and operation to assure that it meets customer requirements. (See preliminary design review, critical design review, and first article configuration inspection.)

Design series, the same as type, model, series.

Detail drawing, a complete description of a single part that includes dimensions, tolerances, finish, materials, and processes required to build, identify, inspect, and test the completed part.

Detail equipment specification, a document that describes the performance, construction, size, reliability, weight, and testing requirements of an equipment. Consists of Part I, design requirements, and Part II, product configuration.

Development, the effort involving design, testing, evaluation, and redesign of a product before acceptance for production.

Development baseline, the baseline established before beginning qualification model fabrication and testing.

Deviation, customer authorization to depart from a specification or other document requirement before the departure is made.

Disposition, the decision to rework, use as is, return to vendor, or discard an item that does not meet specifications or quality standards.

Distribution, the process of issuing the required copies of documents to project members and customer representatives.

Document, the collective term for specifications, drawings, parts lists, standards, and reports. (See data.)

Drawing, a graphic representation of an item.

Drawing levels, the classification of drawings in terms of different applications; for example, in-house construction to low standards or production items of high quality.

Drawing number, a number used to uniquely identify a drawing. Consists of letters, numbers, or a combination of letters and numbers, which may be separated by dashes. (Excludes revision letter.)

Dynamic load analysis, an evaluation of the effects of stresses simulating flight conditions on the operation and survival of a CI.

ECP number, a unique alphanumeric designator assigned to an engineering change proposal.

ECP package, a compilation of data submitted to customer for approval of a change. It includes standard ECP form, change description and drawings, cost estimates, and schedule revisions.

ECP sequence, the schedule of an ECP; for example, before, after, or concurrent with a previous ECP, test point, or assembly level.

ECP sequence number, identifies each ECP prepared during a project. Not related to ECP sequence above.

ECP types indicate the degree of completeness of an engineering change proposal; for example, preliminary is identified with a P and formal is designated by an F, indicating a complete ECP.

Effectiveness, the degree to which a CI can be expected to achieve specific system or mission requirements. Effectiveness is measured in terms of the CI's availability, dependability, and capability.

Effectivity, the approved use of a part or item in specific CI serial numbers.

Emergency ECP, a priority assigned to an ECP because of safety conditions which could cause fatal or serious injury to personnel or severe damage to CI.

End item, a combination of items that form a product that accomplishes a specific task or function.

End item designator, a permanent identifier assigned to the equipment or a major assembly. It may be the CI number or a more concise type designator when the CI number is too long for convenient use in records or computer print-outs where space is limited.

End item identification number, see CI number or end item designator.

Engineering change, any change in design or performance of an item after establishment of its configuration identification.

Engineering change justification code indicates the reason for a Class I engineering change; for example, correction of a deficiency, operational or logistics support, cost reduction, or production stoppage.

Engineering change package, a package of data describing a proposed change. It may include an EO, ECP, drawings, and an SCN.

Engineering change priority, the rank given to Class I engineering changes; for example, emergency, urgent or routine.

Engineering change proposal, a document that describes a Class I engineering change. It is used for proposing a Class I change to the customer for approval before incorporation of the change.

Engineering change request and analysis, see change request.

Engineering critical component, a key component because of its complexity, performance, or critical nature with regard to engineering aspects of the equipment development and production. It is the same as a critical component.

Engineering data, specifications, drawings, parts and wire lists.

Engineering development refers to a CI that is still in the design and evaluation stage and has not been approved for production or operation.

Engineering model, a model of the equipment that is physically, electrically, and functionally the same as the intended flight model but does not necessarily meet construction, parts quality, durability, or workmanship requirements. It is a precusor to the prototype.

Engineering order, a document that describes an approved change to a released engineering drawing, parts list, and so on. The EO is attached to prints of the drawing and is treated as an integral part of the drawing. It is similar to an ADCN.

Equipment, an item designed and built to perform a specific function as a self-contained unit or to perform a function in conjunction with other units. It is the same as a product.

Equipment number, a permanent number assigned by a company to identify a CI. It is the same as CI number.

Exploratory development, an effort to solve specific problems related to a broader goal without conducting a major development project.

Facility, a fixed installation, including RPIE, that is an integral part of a program or system.

First article configuration review (FACR), the same as FACI.

Failure, the inability of an item to perform within previously defined limits.

Failure report describes the failure of an equipment or item, its cause, and the corrective action taken to prevent a recurrence of the failure.

Family designation number, a permanent number that is a part of or used with a part number to establish a base for serializing items that are a part of an equipment. When items in the family become noninterchangeable, the family designation number remains the same, a new suffix number is added, and serialization continues with the previously established sequence.

Federal item identification number, a number sequentially assigned to each item of supply approved by the government.

Federal supply class, a grouping of homogeneous items on the basis of physical or performance characteristics.

Field, the area of a drawing used for description of an item; it includes pictorial representations and notes.

Field change, an engineering change made to a CI officially accepted by the customer. (It bears no relation to field of a drawing.)

Field operation, the use of an equipment at the deployment site, at a facility, or in an aerospace vehicle to accomplish its intended goal or mission.

Field release, the release of engineering data that change formally accepted equipment under the jurisdiction or control of the company and progressing through field testing or operation.

Follow-up report, a report that describes the completion of an approved engineering change to a CI or item.

Find number, a number assigned to an item on the field of a drawing for cross-referencing to a parts list.

First article configuration inspection, a formal review of the as-built configuration of an equipment against its documentation to establish the product configuration baseline for the CI. Formal approval of Part II of the detail specification occurs during FACI.

Fit, the equipment mounting elements, methods, location, and dimensions.

Form, the shape of an equipment.

Form, fit, and function, the physical and functional characteristics of a CI as an entity, but not covering characteristics of the elements making up the CI.

Formal configuration management reviews, preliminary design review, critical design review, and first article configuration inspection.

Function, the purpose or end use of an item.

Functional area, a distinct group of performance requirements that are part of next lower level breakdown of the overall highest level performance requirements of a system.

Functional baseline, the functional configuration identification initially approved by the customer. (See FCI.)

Functional characteristics, performance, operating, and logistics parameters and their tolerances; for example, range, speed, safety, reliability, and maintainability.

Functional configuration audit, a formal examination of functional (performance) test data before customer acceptance to verify that the CI meets the requirements given in its functional or allocated configuration identification. It is similar to CDR.

Functional configuration identification, technical data that give the functional characteristics, demonstration tests, interface characteristics, and design constraints of a product.

General specification, covers requirements common to different types, classes, models, or styles of a product.

Ground support equipment (GSE), a test set used to check out the operation of a product or equipment used for such things as installing, launching, adjusting, repairing, or controlling a product.

Hardware, any item built by the company or a vendor (except software).

Hardware/Software, refers to hardware or software, or both; software applies only to those items associated with hardware for operational use: installation manuals, operating and repair manuals, command and control computer programs, and data recorded on tapes, cards, or discs. The term "hardware/software" does not apply to drawings, parts lists, specifications, manufacturing instructions, and so on.

Higher assembly, an assembly into which an item goes.

History file, a file of obsolete or superseded engineering documents, including change documents.

Human engineering, a design discipline that assures a product can be used safely and properly by people.

Human factors, the application of knowledge about human characteristics to the design of items to achieve effective man-machine integration and use. Also includes personnel selection, training, job performance aids, and performance evaluation. Human engineering is a subclass of human factors.

Identification item, a category of equipment defined as being of simple design and construction.

Identification number, the number used to identify an item. This number may be a specification, drawing, part, model, type, or catalog number, depending on the numbering system of the company. It may be a transient or a permanent number.

Incorporated, an engineering change is incorporated when the drawing vellum is changed.

Indentured drawing list, a complete list of drawings for the equipment. This list is arranged by indentation of the drawing numbers to show the pattern by which drawings go into their next assemblies. Indentures may also be shown by numbers or letters.

In-process tests, production line tests performed at intermediate points between receiving tests and the start of final acceptance testing.

Inseparable assembly, an assembly made of two or more parts that cannot be disassembled without damaging the assembly or its parts.

Integrated logistics support, a grouping of elements for assuring effective and economical support of an equipment at all maintenance levels for its planned life cycle. These elements include (a) facilities, (b) maintenance, (c) support equipment, (d) logistics data, (e) spares and repair parts, (f) logistic support personnel, and (g) contract maintenance.

Integrating contractor, the company responsible for overall scheduling of system checkout of associate contractors and for providing services common to several contractors. Similar to a systems contractor.

Integrating ECP, an ECP priority that specifies and coordinates the impact of a change on all elements of a system, including system requirements, interfaces, and other CEI's.

Integration, the combining of different equipments into a subsystem or system so that they work together harmoniously.

Interchangeable item, an item that is functionally and physically equivalent to and capable of being exchanged with another item without (a) changing either item or adjoining items and (b) selection for fit or performance.

Interconnect diagram defines requirements for interconnecting the various assemblies in an equipment. Data include cable origin and termination, connectors, and terminals.

Interface, a common boundary between two or more items. This boundary may be electrical, mechanical, functional, or contractual.

Interface control drawing (ICD) defines the junction points between the equipment, other equipments, and the system; for example, mounting dimensions and attachments, center of gravity, outside dimensions, weight, and heat dissipation.

Interface directive, a document from the customer's CCB that indicates satisfactory interface coordination and agreement has been reached. It is only required for ECP's having interface aspects.

Interrupt priority, the priority assigned to a break in the normal flow of a system or a routine such that the flow can be resumed from that point at a later time. An interrupt is usually caused by a signal from an external source.

Interrupt alert, a warning indication of an interrupt condition in the normal flow of a system or a routine.

Issuance, the process of printing, stamping, and distributing a document.

Item, a general term applying to piece parts, components, modules, subassemblies, assemblies, equipments, products, and subsystems. The term is not used to refer to a system.

Item identification and part number, the complete number required to identify a part, relate it to its drawing, provide a basis for serialization and for determining the interchangeability or noninterchangeability of the part.

Key card, a control card representing the authorized check sum number used by quality assurance to audit and validate the status of the computer program with respect to its authorized configuration.

Key functional characteristics are critical characteristics that affect the satisfactory fulfillment of an equipment's operational requirements such as range of measurement of an instrument, payload capability of a rocket, or altitude capability of an aircraft.

Kit, a collection of carefully identified and controlled items used to build a module, printed circuit board, subassembly, or assembly. Kit items are usually kept in a plastic box or plastic bag and labeled.

Kit list, a tabulation of all parts and materials that go into a kit. The list includes part and serial numbers and lot numbers or date codes for each item listed.

Layout, the arrangement of items within a module, subassembly, assembly, or equipment.

Level of assembly, the complexity of an assembly. For instance, when a module is attached to a subassembly, the subassembly is a higher level of assembly with respect to the module. Conversely, the module is a lower level of assembly with respect to the subassembly.

Life cycle, the period covering the design, development, manufacture, operation, maintenance, logistics support, and repair of an equipment.

Limited release, a drawing or other document to be used for building an engineering model or other specific hardware not requiring customer acceptance to an established requirement.

Line item, an item called out in a separate line as deliverable to the customer in the work statement or contract.

List of materials, the same thing as a parts list.

Loading, inputing of instructions into the computer so it can perform its required operation.

Log book, a historical record that stays with the equipment and describes all actions related to performance, configuration, and maintenance during the life of the equipment. Failures, rework, replacements, and operating periods are recorded.

Logistics, the same as product support.

Logistics critical component, a vendor-designed, repairable item that may require spares provisioning or repurchase from other sources.

Lot, a collection of production items that have been built from the same material or at about the same time.

Lower assembly, an assembly that is a part of a larger assembly.

Maintenance, the operations involved in keeping an equipment in working order or restoring it to normal operation after a malfunction occurs.

Maintainability, the quality of equipment design and installation that simplifies inspection, test, servicing, and repair with a minimum of time, skill, and resources.

Manual, a document that provides operating and maintenance data for an equipment.

Manufacturer's code, the same as code identification number.

Manufacturing instruction or order, a document that describes the processing, assembly, and inspection steps required to build and accept an item. A separate manufacturing instruction is used for each module, subassembly, and assembly.

Major deviation, a contractual departure from documentation that involves health, performance, interchangeability, reliability, maintainability, repair, effective operation, or weight. (See Deviation.)

Major waiver, a contractual departure from documentation involving health, performance, interchangeability, reliability, maintainability, repair, effective operation, or weight. (See Waiver.)

Master, the same as an original drawing.

Matched pair, two parts that are mated as a pair to provide a function in the equipment. A matched pair is identified with a single part number and cannot be separated without affecting its performance.

Materials (bulk), items used for building an item or equipment the quantities of which cannot be determined; for example, wire, solder, welding rod, epoxy, and insulation sleeving.

Material control, an activity involving identification of materials and fabricated parts, their quantities, and their locations. This is done by placing identifiers on the parts or on records traceable to the fabricated parts.

Minor deviation, a departure from approved documentation that does not involve safety, health, performance, repair, interchangeability, reliability, maintainability, effective operation, or weight. (See Deviation.)

Minor waiver, a departure from approved documentation that does not affect safety, health, performance, interchangeability, reliability, maintainability, effective operation, repair, or weight. (See Waiver.)

Mission, the intended goals or functions of an equipment, system, or spacecraft.

Mission, design, series, the same as type, model, series.

Mockup, a simulated model of the equipment's mechanical or thermal configuration.

Modification, a change to an equipment and spares allowed only after the contract has been revised.

Modification kit, a kit used for changing a production equipment accepted by the customer or released for installation at the launch site or deployment area.

Monodetail drawing, a drawing that only shows one part or item.

Modification work order, a document that provides instructions for modification of an equipment.

Model (equipment model), an equipment that is a member of a similar type of equipment, but which has some differences in performance or physical characteristics.

Module, a prepackaged unit of a subsystem usually treated as a component, such as a pre-amplifier, oscillator, operational amplifier, or gate.

Multiple release, an engineering release record in which superseded and superseding documents are both retained in active files.

Next higher assembly, the assembly into which another assembly goes.

Nomenclature, a name and alphanumeric identifier given to an item or equipment for classification and identification. For example, solar wind spectrometer SWS-1.

Nonconformance report, an inspection report on an item describing its failure to meet its specification, drawing, or quality requirements.

Noninterchangeable item, an item that cannot be used to replace another item without affecting fit, function, or durability.

Nonrecurring costs, costs that result from an engineering change, but which are independent of the quantity of items changed; for example, redesign, qualification testing, or special tooling or jigs.

Nonstandard part, an item that does not meet formal government, industry or company standards or specifications.

Notice of revision, a form for proposing revisions to a drawing, parts list, wire list, or the like. When the NOR is approved, it is used to notify users that the document has been revised. If an SCN is not applicable, a NOR can be used to revise a specification.

Obsolete refers to a document that is no longer used on a project and is not replaced with a superseding document.

Operational applies to actual use of a product.

Operational requirements are functional and performance requirements.

Operational support refers to material, personnel, documents, and items required to maintain the product.

Operational systems development, an effort for developing and testing a system, support program, vehicle, or equipment that has been approved by the customer for production and field operation.

Original drawing, the primary document, maintained by the design activity, depicting an item and containing all official revision data. It is also called a master drawing.

Outline drawing shows external dimensions of equipment and provides special data on connectors and mounting requirements for installation into a system. It is also called a configuration drawing.

Outstanding EO, a design change document that has not been incorporated into the drawing master or original.

Packaging, the total mechanical design of an equipment or item. This includes parts layout, materials used, protective finish, thermal control design, and fabrication techniques.

Part, one or more pieces joined together that cannot be separated without damage to at least one of the pieces. It is identified by a part number.

Parts list, a list of all materials and parts that make up a module, subassembly, assembly, or equipment. A parts list may be a separate list or may be a part of the drawing.

Part number, a number used to uniquely identify a part. It is usually the same as the drawing number, minus revision letter, or includes the drawing number. Its purpose is to control assembly and replacement of items on the basis of interchangeability.

Performance, the functional or operating characteristics of an equipment; for example, measurement range, accuracy, stability, linearity, and reliability.

Performance specification, a specification that describes what is to be accomplished by an equipment but does not describe how the equipment is to be designed.

Physical characteristics, quantitative and qualitative material descriptions of an item; for example, form, fit, dimensions, finishes, and composition. Tolerances for each characteristic are also given.

Physical configuration audit, a formal examination of the as-built configuration of an equipment against its documentation to establish the initial product configuration identification.

Power profile, a graphical representation of the power consumption of an equipment under different operating conditions of a mission.

Prefix, a number or letters added before an identifying number to denote the type of document, such as specification, parts list, wire list, and so on.

Preliminary design review, a formal review of a preliminary design of an equipment to establish compatibility of design, to identify engineering documentation required and to define physical and functional interface relationships between an equipment and other equipments. The Part I specification is approved during this review.

Preliminary technical development plan, the precusor to technical development plan.

Prime equipment, a complex CEI that requires acceptance testing and complete documentation and administrative control.

Print-out, a computer output tape or sheet containing tabulated engineering data.

Priority, see engineering change priority.

Privately developed item, an item completely developed at the company's expense and offered to the customer as a production item. Customer control of the configuration is usually restricted to the item's form, fit, and function.

Process specification, a document that defines the requirements and procedures for performing a process, such as soldering, welding, encapsulating, heat treating, coating, and plating.

Procurement, the purchase of parts or materials for the equipment.

Procuring agency, the government organization responsible for issuing a contract for work to be done or for an equipment to be delivered.

Producibility refers to the ease of manufacture and assembly of an item, including access to its parts, tooling requirements, and realistic tolerances.

Product refers to a system, equipment, component, data, or operational computer software.

Product allocation list, gives location of each CI and the quantity requirements for each location or system.

Product baseline, the product configuration identification initially approved by the customer.

Product configuration baseline, a CI baseline defined by an approved Part II of the detail equipment specification, which is established by completion of FACI.

Product configuration identification, the customer-approved technical data that defines the equipment's configuration during production, operation, maintenance, and logistics support phases of its life cycle. The PCI usually prescribes (a) required physical characteristics of the CI; (b) selected functional characteristics specified for acceptance testing; and (c) production acceptance tests to be conducted. It is the same as Part II of the detail equipment specification and related engineering data.

Product equipment baseline, the baseline established before beginning production to which the configuration item manufacture is controlled. This baseline is the basis for control during the production and operational period. It is represented by Part II of the detail equipment specification.

Product specification, a document that gives procurement, production, and acceptance characteristics for a production item below the system level.

Product support, the project area responsible for supplying spares, instruction manuals, training facilities, and modification kits. It is also responsible for transportation of equipment and its maintenance.

Production, the manufacture of formally accepted products beginning with FACI.

Production baseline, a company baseline that precedes the customer product baseline.

Production release drawing, a drawing for manufacture or procurement of items specified for use in the CI's.

Program, a related series of efforts to attain a broad scientific or technical goal. Also used in a specialized sense as follows: reliability program, quality assurance program, test program, and computer program.

Program (computer), a means of directing and controlling the operation and functions of a digital or analog computer.

Program generator, an on-call coded sequence of instructions for a computer process.

Program specification establishes requirements for a related series of projects designed to accomplish a broad scientific or technical goal.

Project, an effort within a program that may involve research and development, design, construction, and operation of an equipment.

Proofing requirement, the need to demonstrate that a CI meets form, fit, and function requirements.

Protective treatment, a treatment of the equipment to prevent corrosion or fungus development.

Prototype, the first model of the equipment that meets all functional and physical requirements and is thus identical to the production equipment, with minor exceptions as specified in the contract. Its purpose is to demonstrate that the company is ready to begin production.

Provisioning, the act of providing spare parts for an equipment.

Qualification, the determination by tests, examination, documents, and processes that an item meets or exceeds the expected environmental stresses without failure or malfunction.

Qualification model, the CI used to demonstrate that the basic design and construction can withstand the maximum stresses expected during operation.

Qualification status list, identifies each item that goes into the equipment and its source or status of qualification for use.

Qualified products list, an official list of approved items for use on a project or program. It is usually issued by the customer.

Quality assurance, the system, procedures, and activities for assuring that an item will perform satisfactorily in actual operation by verifying that materials, construction, and testing meet the requirements of project and procurement specifications.

Range safety, refers to considerations related to the safety of personnel and nearby populated areas during launching of a spacecraft or sounding rocket.

Record change, a Class II change that only affects a drawing and does not affect materials, parts, tools, costs, or schedules.

Recurring costs, costs that will be incurred for each item built or document prepared during the project. (See nonrecurring costs.)

Reference release drawing, a document released for information only and not used for building, inspecting, or testing.

Registration number, a number assigned by the government to an equipment to indicate government ownership, responsibility, and accountability.

Release, the process of furnishing a document to data control and authorizing its issuance to manufacturing. Only engineering is allowed to release a document to data control.

Released engineering design, the current and total set of drawings and specifications for an equipment that has been completed, recorded, and made available for manufacturing or procurement.

Release record, a record that contains key data (such as data of issue, number, revision letter, and originator) on a document released by engineering for information or for fabrication, test, or procurement of an item.

Reliability, the probability that an item will perform its function for a specified period in its intended environment.

Reliability prediction, a forecast or estimate of the probability that an equipment will operate satisfactorily for a specified period under the expected environment.

Repair parts, piece parts or nonreparable assemblies for repairing spares or major end items.

Replaceable item, an item that is functionally interchangeable with another item, but which differs physically from the original item, making necessary special operations (filing, drilling, reaming, and so on) before it can be installed.

Retrofit, the incorporation of an engineering change in an equipment accepted by the customer or in service. (Retrofit is also referred to as a service action.)

Retrofit kit, see modification kit.

Revision letter, a letter added to a drawing or other document number to indicate that an engineering or other type of change was made to the original document.

Revision, a modification of a drawing or design after it has been officially released by engineering.

RFI, the radio frequency interference produced by the equipment or by other equipment in the system.

Routine ECP, the priority given to ECP's that cannot be assigned a compatibility, emergency, or urgent priority.

Real property installed equipment, a customer-owned item that is physically attached to, integrated into, or built into the configuration item.

Schedule, a statement of times for projected tasks and events.

Scratch program, an auxiliary program representing the master file of a computer program, which is used for trial and test conditions for proposed computer program changes.

Screened component, an item subjected to operating tests before CI installation or operation.

Seller, the same as company or contractor.

Sequence number, the address or order of a data item stored in the computer or in a deck of punched cards.

Serial number, an identifier used with a part number to denote each unit in a family of similar items. It provides for effectivity identification of design changes.

Series designator, an identification element within a model for CI's that have the same basic design, but not necessarily identical configurations.

Shop order, the same as a work order or manufacturing instruction.

Source, see supplier.

Source code, identification code of the manufacturer of an item.

Source document, a primary engineering document, (drawings, specifications, and so on).

Source control drawing, a drawing that specifies configuration, design, and test requirements for a commercially available item (not military standard items) and the sources or vendors that are exclusively qualified to supply the items. An SCD is necessary when a critical item has demonstrated its satisfactory operation in a CI but other items have not demonstrated their reliability and performance under the same conditions.

Spare, a component or assembly of an equipment held in reserve to be used when an in-use item fails to operate satisfactorily.

Specific operational requirement, a document which describes operational or performance characteristics needed to fulfill a near-term operational requirement for a system.

Specification, a document, primarily used for procurement (purchase of an item from a vendor or subcontractor), that describes the major technical requirements for an item and the procedure for determining the requirements have been met. Key sources of specifications are the Federal Government, the military, and industry.

Specification change notice, a document that describes changes to an approved specification. The SCN is made a part of the specification after customer approval.

Specification control drawing, ensures that a vendor-supplied item meets minimum requirements and is functionally and physically interchangeable with items previously ordered by the same part number. An SCD is necessary to prevent vendors from significantly redesigning the part without changing its part number.

Specification identification index, a record of all equipment specifications and approved changes for an equipment or system.

Specification maintenance, the procedures or process of changing a specification to incorporate approved changes and corrections to the text, including preparing documents for recording authorizations and history of changes and revisions.

Specification tree, a drawing showing the indentured relationships among specifications independent of the assembly or installation relationships of the items specified. The tree shows the dependency of specifications on other specifications.

Standard, a document designed for recurring use. It specifies engineering and technical limitations and applications for an item, process, or engineering practice. A standard gives general requirements and does not describe how something shall be done. Key types of standards are federal, military, and industrial.

Standard configuration identifiers, the complete set of numbers used to identify the configuration of the equipment and its spares. This set consists of six numbers: (a) specification identification number, (b) CI number, (c) serial number, (d) item identification and part number, (e) change identification number, and (f) code identification number.

Standard part, an item that is identified by a federal stock number, specified by a military standard, a source control drawing, or by other specifications prepared by a contractor.

Static load analysis, an evaluation of the effects of anticipated steady-state loads on equipment operation and reliability.

Stock number, a federal identification number for an equipment specified in a contract by a government agency for inventory control.

Stop order, a cease work order used when fabrication or procurement of a specific item will be affected by an impending change, resulting in expensive rework or replacement.

Storeroom, a room for storing parts and materials for the equipment. Only authorized project members have access to this room if it is a controlled or bonded storeroom.

Stress, the mechanical or electrical forces that are applied to an item or equipment during testing or use.

Subassembly, two or more parts that form a portion of an assembly replaceable as a whole but having a part or parts that are individually replaceable.

Subcontractor, one who performs a subtask for the company that has the equipment contract.

Substitute item, an item that can be used to replace another item but only under certain conditions or applications. It is functionally and physically interchangeable and does not require any changes to be installed.

Subsystem, a major functional subassembly or group of items that is essential to operational completeness of a system.

Suffix number, a number added after a part number to denote noninterchangeability.

Superseded, refers to a drawing or other controlling document that is replaced by a new document.

Supplier, the same as vendor and subcontractor. A supplier is a company that provides a service or produces an item for the equipment.

Support equipment, equipment required to make the CI operational in its intended environment; for example, ground equipment or computer programs.

Synthetic part number, a temporary, unofficial number assigned to a part by the manufacturing group to simplify processing and handling.

System, a grouping of subassemblies, assemblies, and equipments that provide a specific function. A system may also include personnel and facilities.

System analysis, the systematic approach to trade-off studies and the development of more complete specifications.

System effectiveness, see effectiveness.

System configuration chart, a record of approved ECP's and revisions to a system specification.

System designation number (or system designator), an alphanumeric identifier, such as LSEP 606, for a system family (or type, model, series) that will include the equipment to be supplied by the company.

System engineering, the engineering management of a total system to determine and maintain technical integrity over all system elements. This includes system design approach, system synthesis, system analysis, functional analysis, requirements analysis, and task analysis.

System specification, a general specification containing technical and mission requirements for the system as a whole and apportioning these requirements to subsystems or equipments for meeting mission goals. It also defines interfaces between the different items.

Tabulated drawing, a drawing showing similar items which as a group have constant and variable characteristics. Each item is identified by a dash number and by a description of its variable characteristics. The constant characteristics are given in the drawing (depicted graphically).

Technical development plan, a complete description of the effort required to fulfill a need, including identification of high risk areas, functional diagrams, equipment configuration, gross solutions to system requirements, and funding schedules.

Technical manual, a type of technical order which contains instructions designed to meet the needs of personnel engaged in operating, maintaining, servicing, overhauling, installing, or inspecting the equipment.

Technical order, an official document describing technical information, instructions, and safety procedures related to operation, maintenance, installation, or modification of an equipment.

Technical requirement, a condition given in a specification that must be satisfied by the product.

Teletypewriter exchange, a printed message that is produced by a teletype machine.

Time compliance technical order, a type of technical order that gives instructions (sometimes with time limitations) for modifying a product, performing or establishing special inspections, or imposing temporary flight restrictions.

Time critical (time limited) component, an item with a finite life, which, if not monitored and replaced after a prescribed time, could result in a product failure.

Tolerance requirements studies, analyses of mechanical or electrical tolerances required to achieve a desired goal.

Tooling, jigs, and fixtures, the hardware used to build, adjust, or test an item.

Top assembly drawing, the highest drawing describing the product or CI purchased by the customer. It shows three views of the CI and includes a list of major parts, an identification plate, and attaching hardware.

Traceability, the ability to determine the origin and date of manufacture of a part assembled into a product or to determine which serial numbered product contains a part from an identifiable lot.

Trade-off, an evaluation of a design change to determine its importance in regard to benefits versus disadvantages (higher cost, delays, and so on).

Training equipment, the maintenance and operator training aids and related software used for training operating and maintenance personnel.

Transient number, a nonpermanent number used for internal company control or temporary identification of an item.

Transportability, a design characteristic related to the building of a CI so that it can be transported to its destination. Considerations include weight, size, dangerous features, and sensitivity to shock, vibration, temperature, and humidity.

Type, the basic classification of an equipment; for example, radiation detection, telemetry, data processing, radar, meteorological, or television.

Type designator, the alphanumeric identifier for shorthand identification of an equipment based on function or use.

Type, model, series, a quantity of equipments of one basic design that is specified by one detail specification as a block of items to be designed, developed, and built to that specification. The block of equipments is identified by the CI number assigned by the company or customer.

Type number, the same as type designator.

Uniform specification program (USP), the preparation, approval, and maintenance of specifications required for configuration management. Specifications include system, detail equipment, component, and standard documents.

Unit, an assembly or combination of items capable of independent operation under a variety of situations; for example, an aircraft, spacecraft, transmitter, radio, electric motor, and power supply. It is similar to the terms "equipment" and "item" but includes systems and excludes subassemblies and piece parts.

Urgent ECP, an ECP priority for design changes affecting (a) equipment safety, (b) schedule, (c) mission capability, and (d) interfaces with other equipments that have to be changed.

Value engineering, a discipline involving the analysis of a design to determine how the proposed equipment can be produced more cheaply by eliminating unnecessarily complex design and expensive materials or parts without affecting the equipment's overall performance.

Vellum, a translucent sheet of paper that contains the original drawing information and subsequent revision data. (Same as original drawing.)

Vendor, a manufacturer or supplier of a commercial item.

Verified configuration record, a document that identifies the serial numbers and drawings (including revision letters) to which each module and subassembly has been built for a particular equipment. (Identical to as-built list.)

Waiver, the customer acceptance document for an item that does not meet a specific contractual requirement.

Weight analysis, an itemized list of parts, subassemblies, materials, and chassis weights for an equipment or assembly.

Work breakdown structure (WBS), a product-oriented family tree composed of hardware, software, services, and other tasks; the WBS results from project engineering effort during the development and production of an item and completely defines the project. A WBS displays and defines the product to be developed or produced and relates the elements of work to be accomplished to each other and to the end products.

Work order, a written authorization and instruction to perform a task.

Work statement, a detailed description defining tasks to be done, equipment and documentation to be delivered, delivery dates, performance requirements, applicable specifications, and standards. It is included in contract by reference.

Working deck information, a summary of information comprising the working file of processing instructions and resident information in the computer memory.

Zone, a combination of numbers and letters around the border of a drawing for identifying the location of a change on a drawing. The letters increase vertically up the drawing and numbers increase from right to left.

INDEX